MAPMAKING WITH CHILDREN
SENSE OF PLACE EDUCATION FOR THE ELEMENTARY YEARS

DAVID SOBEL

HEINEMANN • PORTSMOUTH, NH

Heinemann
A division of Reed Elsevier Inc.
361 Hanover Street
Portsmouth, NH 03801-3912
http://www.heinemann.com

Offices and agents throughout the world

The author and publisher thank those who generously gave permission to reprint borrowed material.

Figure 7–1 from *My Father's Dragon* by Ruth Stiles Gannett. Copyright © 1987. Reprinted by permission of the publisher, Random House, Inc.

Figure 7–2 from *The Ghost of Lost Island* by Liza Ketchum Murrow. Copyright © 1991. Reprinted by permission of the publisher, Holiday House.

Library of Congress Cataloging-in-Publication Data

Sobel, David, 1949–
    Mapmaking with children: sense of place education for the elementary years / David Sobel.
        p.   cm.
    Includes bibliographical references.
    ISBN 0-325-00042-5
    1. Map drawing.   I. Title.
    GA130.S6   1998
    372.89'1—dc21   98–14641
                                                          CIP

Editor: William Varner
Production: Melissa L. Inglis
Cover design: Linda Knowles
Cover artwork: Craig Altobell
Manufacturing: Courtney Ordway

Pen and ink drawings throughout the book are by Cynthia Mathewson.

Printed in the United States of America on acid-free paper
02   01   00   99   98   RRD   1   2   3   4   5

Across Devon's foggy moors
We bounced and squished

On Truro dunes
We tumbled and rolled

Through San José's traffic
We twisted and turned

In Monadnock's shadow
We treasure hunted

This book is dedicated to Eli, Tara, and Wendy—
my true blue exploring companions

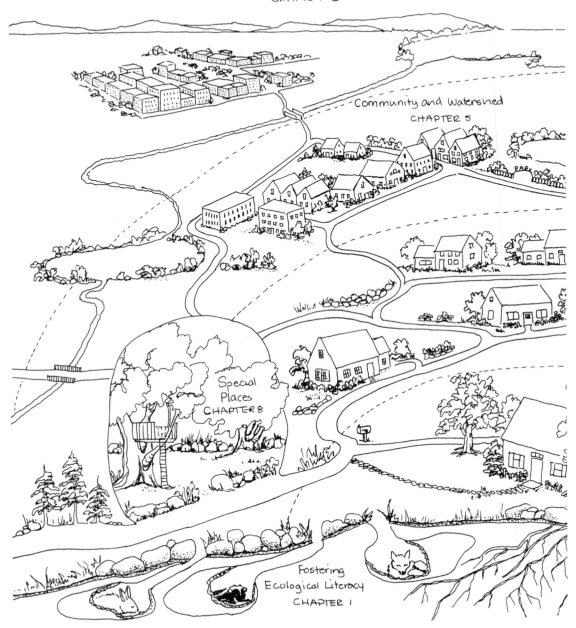

Nation, Region, and Beyond
CHAPTER 6

Community and Watershed
CHAPTER 5

Special Places
CHAPTER 8

Fostering Ecological Literacy
CHAPTER 1

As the child's world expands, so should the curriculum. This visual table of contents illustrates the relationship between the structure of the book and the child's evolving connection to place during the elementary school years.

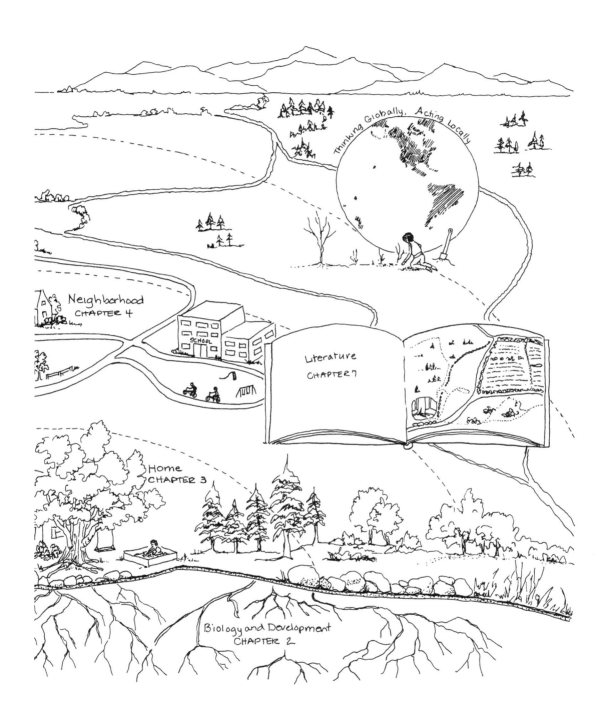

Thinking Globally, Acting Locally

Neighborhood
CHAPTER 4

SCHOOL

Literature
CHAPTER 7

Home
CHAPTER 3

Biology and Development
CHAPTER 2

CONTENTS

EXPLORING HIDDEN LANDSCAPES

ACKNOWLEDGMENTS

In the beginning of this journey, I had a sense of my destination, but I didn't really have a map. Luckily, there were many folks along the way to show me the terrain, provide directions, and travel along with me for a while. My thanks to all the people who provided a home away from home for me and touched this work in some way.

I started this project more than ten years ago while doing research with children and teachers in Devon, England. Though I got sidetracked and wound up writing a different book, my experiences in the tucked away villages surrounding Dartmoor are the foundation of much of this work. Alan Forster, then Warden of the Newton Abbott Teacher's Center, got me oriented and provided wise guidance. Len Peach, Ann Strut, Pam Gordon, and Nick Buchanan, the teachers at the Denbury Primary School, graciously shared their students, time, and tea with me and made me feel like one of the staff as I conducted mapmaking research with each child in the school. Rosemary Riddell of the South Brent Primary School collaborated with me on a wonderful Parish Maps project in her classroom.

In northern New England, I recruited a group of thoughtful and developmentally sensitive teachers to document the fine work in social studies, geography, and literature that they were conducting in their classrooms. Many of the approaches, children's work, and classroom profiles emerged out of their project-centered curricula. My appreciation and thanks to: Craig Altobell, Cogsbill Middle School, Henniker, NH; Deanna Avery, Nantucket Elementary School, Nantucket, MA; Heather Barber, Blue Valley School, San José, Costa Rica; Amy Carter, The Harrisville Children's Center, Harrisville, NH; Julie Dolan, Townshend Elementary School, Townshend, VT; Mickey Johnson, Temple Elementary School, Temple, NH; David Millstone, Marion Cross Elementary School, Norwich, VT; Terry Monette, Harvard Elementary School, Harvard, MA; Steve Moore, Park Street School, Springfield, VT; Mary Morrisette, Chesterfield School, Chesterfield, NH; Maggie Stier, Friends of the John Hay Estate, Newbury, NH; Heather Taylor, The Neighborhood Schoolhouse, Brattleboro, VT; Maureen Woolford, Merrimack Elementary Schools, Merrimack, NH.

Cynthia Mathewson has made a significant contribution to making the book beautiful and accessible. She worked with me to translate the abstract ideas into visual form and patiently worked and reworked many images until we were both happy with them.

We were both particularly satisfied with the visual table of contents. This idea was least well formed in the beginning, but compelling to both of us. Cynthia's grasp of the core ideas of the book and her understanding of the child's perspective on the natural world really comes through in this unique image that conveys the core themes of the book on one page.

Craig Altobell's collage map for the cover was based on an activity he has done with his fifth- and sixth-grade students while doing a place-based study of the Contoocook River in New Hampshire. He has students create a variety of painted papers using many different techniques—watercolor washes, bubble printing, texture imprints, and so on. After numerous river field trips, he asks students to write haiku about their experiences and then illustrate the haiku by making collages from the painted papers they have made. The results are elegantly beautiful; the process seems to help students capture the balance between geographic and poetic knowing of the landscape. I appreciate his unique blend of fine teaching and artistry.

At Antioch New England Graduate School, many of my colleagues have contributed in some way. I have particularly enjoyed my collaboration with Julie King, with whom I have taught Integrating Math and Science Through Mapmaking. Many of the ideas in this book were articulated and refined during our work in this course. I appreciated the students in this course who willingly served as guinea pigs and produced some of the beautiful maps included in the book. I have also enjoyed my work with Paula Denton on a pair of courses entitled The Ecology of Imagination in Childhood and The Study of Place. Many useful ideas have emerged from the integration of these courses.

Finally, I am indebted to Caroline Wiggin, the office manager for the Education Department, who helped with innumerable challenges along the way. And my sincere thanks to all my colleagues in the Education and Environmental Studies departments who pitched in to cover my responsibilities while I took a sabbatical to complete much of the writing. It takes a village to write a book.

1 | FOSTERING ECOLOGICAL LITERACY THROUGH MAPMAKING

THE LURE OF EXPLORATION

It was our heart's desire the whole autumn that I was in fourth grade. The corn fields, overgrown pastures, thickets, and wetlands of the old farm stretched out just behind my friend Kevin's house. They were strictly off limits, of course. The No Trespassing signs hung on the barbed wire at the far end of our kickball left field. But this patchwork of countryside lured Kevin and me as if we were compass needles and there was a giant horseshoe magnet buried in a fieldstone root cellar out there.

Green's Farms, our little corner of the Connecticut coastline, had undergone its first phase of suburbanization by the mid-1950s. The shoreline was mostly claimed by elegant seaside mansions, but the interior was still a mix of scruffy wild places and casually tended farms. We had an array of haunts including the salt marshes and phragmites thickets of Sherwood Island State Park, the tidal flats, and the haunted house—lair of hoboes and ghosts—with its decaying outbuildings and greenbriar thickets. We explored the railroad station, the railroad tracks lined with gravel pits, honeysuckle, and seaside rose, and the sea walls, pocket beaches, and immense rock jetties of the fancy houses. In the summer our adventures were all shore based, but in the fall we became more stealthy and secretive. We headed for the interior (see Figure 1–1).

It started with a series of reconnaissance missions, short forays into the wilderness. After cookies and milk at Kevin's we would head out into the yard to play and, when no one was watching, surreptitiously slip under the bottom strand of barbed wire. There were rumors of shotguns loaded with rock salt, and mean guard dogs, so a fever pitch of alertness always prevailed. First we would make it to the edge of the nearest corn field and then retreat. The next time we'd make it down the dirt road into the woods to the old garage. After each exploration we returned to Kevin's room to review our new discoveries. "The dirt road leaves the first corn field from this corner. There are really four corn fields, not three. Did you see the bats fly out the windows when we opened the garage door!? I wonder where the road goes after the garage?" When it became too much to keep in our heads, we drew a map, hid it under Kevin's bed, and revised it after each foray.

It was not until November that we discovered the water tower, deep inside the property and located next to a puzzling, perfectly round pond with a perfectly round island in its center. The tower became the central feature of the map as we worked to figure out all the different possible access and escape routes. "How will we get away if he drives up the gravel road? Will he see us as we cut through this field?" And when we crept inside the tower and saw the ladders that climbed up and up and up to a small windowed platform a hundred feet above, we knew our fate was sealed. If we could make it up there, we'd be able to see everything sprawled out around us.

Making progress up the ladders was like inching up an ice and rock pitch in the Himalayas. The ladders were vertical, creaky, and covered with pigeon droppings. We would make it to the middle of the second ladder, imagine we heard a car coming, and then run the half mile or more back to Kevin's house, never stopping to look back. We'd call each other chicken to taunt ourselves to climb higher, but fear would overcome us. After four trips and a bout of sewing machine legs, the platform and view were ours. The chimney of Kevin's house was barely visible beyond the maples bordering the fields. We could see out to Long Island Sound, to my house on the low ridge next to Burying Hill Beach, and over beyond the girls' school

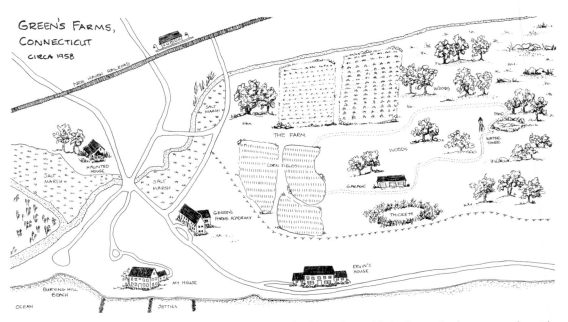

FIGURE 1–1 *Green's Farms, Connecticut, circa 1958. An explorable and mapable landscape for the autumn of my 4th grade year.*

down to the salt marshes. Even better, in the other direction, was an unex-pected expanse of woods and freshwater wetlands that stretched all the way to the beach in Southport. "That stream must flow into the upper end of the salt marsh. I wonder where it starts?" New horizons to explore and map. This was my land, from sea to shining sea.

Somehow, we never continued the explorations after that fall. Kevin and I drifted apart, and the map got lost or thrown away. I remember fi-nally making it to the perfectly round island one winter across the frozen pond, but the unexplored territory beyond remained unclaimed. It's funny how it still gnaws at me, how I still want to feel in my body how it all fits together, to put the last pieces into the puzzle.

I escaped from crowded suburbia after high school and settled into the comfortable wilderness of southern New Hampshire, but a piece of me still feels rooted in the Connecticut coast. The sweet and sour smell of honeysuckle mixed with the sulfury smell of low tide makes me feel at home. The feel of sun-softened asphalt under my feet evokes the thrill of night explorations. Whenever I draw a new map, I feel echoes of those first attempts to make paper match place. That first map was our way of both getting deeper into and stepping back from our discoveries—it preserved what we knew and launched us into further adventures.

And I get excited when children share their homemade maps with me. In these maps I see their active yearning to make sense of their nearby worlds, their desire to record and share their own discoveries and their sense of connection to place. "Here's the kick-the-can hiding place. That's the little path to Erin's house. The cross is where we buried our cat Noah." The stories of their lives are folded into the niches of their neighborhoods; their maps are the weaving together of inner emotion and external forays. Maps are the clothespins that hitch our lives to our places.

PATHWAYS INTO THE WOODS

Let us get one thing straight from the beginning. This book is about my conviction that mapmaking is a crucially valuable tool in elementary schools. I think mapmaking is an inherent human endeavor. Mapmaking, in the broad sense of the word, is as important to making us human as mu-sic, language, art, and mathematics. Just as the young child has an innate tendency to learn to speak and count and sing and draw, the child also tends to make maps. Tony Kallett (1995) lays out this idea in a wonderful little article entitled "Homo Cartographicus." He says,

> It seems to me that one can think of mapmaking as a fundamental human activity, if not *the* fundamental human activity. . . . Learn-ing consists of looking at something new and beginning to see paths into it. You construct a map or a series of maps, each one an approximation and probably wrong in many details, but each one

helping you to go further into the territory. . . . We all have hundreds, thousands of maps each of which represents a way we have learned to look at part of the world . . . There are music maps, language maps, maps of social relations, maps of the physical environment. . . . What they have in common is that all of them are models in our minds of what we think the world looks like and we can consult them to help predict what the world is going to be like, what the consequences of our actions are likely to be.

Kallett's description of learning captures perfectly the experience Kevin and I had exploring the old farm. We followed paths into it and used the map to assemble our experience. In the beginning the map was woefully incomplete, but it was as much as we knew at that point, it was ours, and it helped us go further.

Many teachers will recognize this as analogous to the writing process approach to teaching reading and writing. Children begin the reading and writing process by telling stories, drawing a picture of their story, and then writing—in their own words and with invented spelling—about their picture. The sketchy picture and meager words only slimly approximate the real thing. But they are beginnings, first steps down the pathway into the landscapes of drawing and writing. In spelling, children often first learn to spell their own name, then *Mommy* and *Daddy*, then *I Love You*. These first words are a clearing in the woods, a known place they can come back to and use as a reference point for figuring out the way to begin spelling other words.

I like this broad metaphoric use of the notion of mapmaking. Some educational theorists and neuropsychologists refer to constructing these maps of understanding as *conceptual mapping*. The term suggests less of a linear and sequential model for how we organize knowledge in our brains and more of a spatial, multidimensional process. It's like the difference between playing tic-tac-toe on a napkin and playing the three-dimensional version. In British schools, educators add the skill of graphicacy to the traditional objectives of literacy and numeracy. By *graphicacy* they mean the ability to produce or comprehend visual representations of information, such as drawing, creating collages, constructing graphs, making diagrams, and mapmaking. According to the Inner London Education Authority (1986), "The understanding of simple road signs, complex wiring diagrams, geological cross-sections, house plans and topographical maps all involve the skills of graphicacy."

The concept of graphicacy forms a backdrop for the main thrust of the book. Much good work has been done recently on the value of using concept mapping as an instructional device and as a tool for helping children organize their own thinking. An increased emphasis on understanding visual information is part of the National Council for the Teaching of Mathematics (NCTM) standards for mathematics education and the American Association for the Advancement of Science benchmarks for science education. From an evolutionary perspective, maps are one of the earliest forms of visual information and are inherently accessible to children. Maps are a valuable bridge between the real world and the abstract world and can prepare children for understanding graphs of mathematical data and tables of scientific information. From a cognitive perspective, an increased emphasis on mapmaking will enhance the objectives of greater mathematical and scientific literacy.

My emphasis in this book, however, will be to balance affective *and* cognitive approaches. Mapmaking is useful for the purposes of teaching the content of the social studies and geography curricula, and it also serves as a tool for developing a sense of place. The approach I advocate roots the cartographic experience in our visual, kinesthetic, and emotional experiences.

We do a disservice to children when we jump in too quickly at a prematurely abstract level in map reading and mapmaking. It's important to have children begin mapmaking the way they begin drawing; maps and drawings are representations of things that are emotionally important to children. In the beginning, children's maps represent their experiences of beauty, secrecy, adventure, and comfort. With these affective endeavors as a foundation, I then gradually start to focus on scale, location, direction, and geographic relationships. The development of emotional bonds *and* cognitive skills needs to go hand in hand in my approach to developmentally appropriate social studies and geography.

FROM THE OUTSIDE IN:
PUTTING THE CART BEFORE THE HORSE

This process of working from the inside out, of helping children construct their own maps of understanding, has contributed to the revitalization of many curricular approaches in the last two decades. *Math Their Way* has built on children's naturalistic ways of understanding numbers, what Howard Gardner (1989) refers to as "digital mapping." Science educators talk about the need to have children articulate their intuitive conceptions, wrong though they may be, as a first step to building sturdy concepts. And despite the conservative backlash against the whole language approach, I firmly believe that it genuinely develops real literacy in our children. (Critics of whole language say it decreases reading competence; I say the real culprit is television.) But in many places, social studies and geography education have not been reborn with this same developmentally appropriate vigor. Just as the whole language approach supports reading and writing from the inside out, I want to advocate for a holistic approach to the teaching of mapmaking. I refer to it as the "small world" approach.

You have probably heard about the crisis in geographic education in the United States. Fifteen percent of our fourth-grade students can't find the United States on a world map (Mitchell 1991), 55 percent don't know the capital of France, and so on. This crisis has brought on an array of new programs in geographic education. Some are recitation- and drill-oriented with renewed emphasis on memorizing the state capitals and on geography bees. The National Geographic Society's Geography Education Program is an outside-in curriculum approach with an emphasis on the five major themes of geography: location, place, human/environment interactions, movement, and regions. The problem with this approach is that it denies children firsthand experience. The program emphasizes abstract, long ago and faraway information instead of focusing on the here and now of the child's world. This leads to children who have lots of facts and little understanding.

My seven-year-old son, Eli, has picked up on my love of maps, so we spend a lot of time looking at them and he will voluntarily draw maps for me frequently. His favorites are ski area maps, panoramic rather than aerial view maps of all the lifts and trails at the mountains we visit. He is just starting to be able to make sense of these. While he was in first grade he drew two maps for me—one of the neighborhood we were living in during our four-month stay in Costa Rica and the other a map of the route from home to his school. The neighborhood map was a bit convoluted, but there were many recognizable elements and it showed some correct spatial relationships. His map of the route from home to school included a couple of landmarks in their correct sequential order, but the school wound up being located right behind the gate to our neighborhood. In actuality, they were about 15 kilometers apart. These maps are like the map Kevin and I drew of the old farm—incomplete and inaccurate, but good tries at making sense of experience. These are appropriate challenges for my son to be working at.

On the other hand, Eli came home from his first-grade class during that same time proudly displaying his book of continents—a perfect example of the outside-in approach that I am not enthusiastic about. Following the teacher's instructions, he had dutifully traced around the prefabricated continent shapes and then colored them in messily. South America was all red, Antarctica had some blue splotches on it, Australia looked like a blue-and-brown zebra. He was very proud of it, mostly because he knows I like maps, but he had no idea at all what continents were or which continent we were on. Asking first graders to make maps of their neighborhoods makes sense; asking them to make maps of the continents puts the cart before the horse.

Another example. I was surprised last year to learn that the first-through third-grade multiage class at our local public school was studying the solar system. I have always been puzzled by the curricular commitment to studying the solar system. There is barely anything you can do that is tangible and hands-on with the solar system; very few teachers actually have night sessions so children can at least look at the planets and the moon. Instead, everyone makes scale models of all of the planets, and very few children gain any significant understanding. In the upper elementary grades studying the solar system jibes with a developmental interest in exploration and outer space, but its presence in the early grades seems frivolous to me. When I asked the teacher why she was doing it, she said, "It's in the district science curriculum. I have to teach it."

Regardless of my concerns, Monica, the first-grade daughter of friends of ours, loved the unit and knew all about the rings of Saturn and how hot it was on Mercury. She could breezily recite the order of the planets from the sun out to Pluto and even knew the name of a couple of the moons of Jupiter. On her way to a winter vacation with her parents, however, Monica asked, "Mommy, which planet is Mexico on?" Monica's ability to recite the names of the planets does not mean that she had a grasp of planetary geography or that she had developed any sense of scale.

These outside-in approaches to teaching do not do much to further the goals of geographic education. In actuality, they may do just the opposite. Instead of connecting children to place, this approach alienates them and cuts them off from their local environments. The inadvertent hidden message is: *Important things are far away and disconnected from children; nearby things, the local community and environment, are unimportant and negligible.* Learning is copying someone else's shapes and consuming someone else's facts; learning isn't about drawing your own maps and finding out things for yourself. Does this alienation from local places lead to roadsides littered with discarded trash?

Don't get me wrong about having students engage with content. I knew all the state capitals by the time I was nine and I loved coloring in all those maps of Europe and Africa. And I quiz my own kids about the capitals of the United States. These approaches are fine when they are not done in isolation. Optimally, teachers will use both inside-out and outside-in approaches to mapmaking and social studies education in their classes, and at times the two approaches will converge.

Craig Altobell, while teaching fifth and sixth grade in Henniker, New Hampshire, solved the map of the solar system problem in an ingenious fashion. He asked the students to assume that the distance from the sun to Pluto is a mile. Then he took the class for a walk using a pedometer to figure out exactly how far a mile was and what landmarks there were along the way. When they get back to the classroom, the students make maps from the school to the mile-away point, showing all the landmarks along the way. Then, after they've done the math, they lay the planets out along the route they walked. The sun is right in front of the school, and most of the planets are right there in the school parking lot. But it's a long haul from Jupiter out to Uranus and Pluto. Mapping the structure of the solar system onto a neighborhood map honors the students' relationships with the community and provides an elegant bridge between the known and the unknown. This is the kind of developmentally appropriate geography that we need more of. (See Chapter 6 for a fuller explanation of this project.)

The Brookwood School in Manchester, Massachusetts, has recently revised its science curriculum with a focus on aquatic environments to take advantage of the range of watery places accessible from the school. Starting with a focus on woodland streams in first grade, the curriculum moves down the watershed to ponds in second grade, freshwater wetlands in third grade, and eventually out to the ocean by eighth grade. The first graders' streams are right outside the science classroom's door, the ponds are a bit of a walk, and the ocean is half a mile away. Thus the curriculum expands outward along with the scope of the children's interest and capabilities.

A curriculum based on building a relationship between the structure of the local landscape and the shape of the children's lives must replace our nonsensical focus on the long ago and faraway. We need a curriculum that aspires to ecological literacy—a deep understanding of the flora, fauna, water, culture, climate, and communities that children live in. Whether the class is in the hills of New Hampshire or the boroughs of New York, the initial emphasis should be on what is right outside the door. Lucy Sprague Mitchell, author of *Young Geographers* (1991) and a teacher at the Bank Street School, focused all of her projects for primary children on Manhattan. Children roamed the docks, fish markets, ethnic neighborhoods, and construction sites of the city and made maps of them. When her children outgrew the immediate environment in their interests and needs in the upper elementary grades, the movement outward was organic:

> When the children first leave New York City and the immediate environment can no longer supply all the source material, they will probably follow the routing of some produce they have seen arriving in the city. Usually this will take them up the Hudson by boat or train. Or they may be tracing their water back to its source in the Catskills. (Mitchell 1991)

The expansion beyond the local environment involves tracing one of the pieces of the web of interconnections that ties city dwellers to the ecology of the surrounding countryside.

At the Greenfield Center School in Greenfield, Massachusetts, the curriculum expands outward in a similar incremental fashion. Kindergartners make block models of the classroom. After field trips to the furnace room, the office, and all of the classrooms, first graders model the whole school. In second grade, students sometimes look at the heating system in the school. They locate and map where the oil is stored in the school, who delivers it, how the heated air gets to the classroom, where the oil company's storage tanks are, and how the oil gets to them. By third grade the students study the city of Greenfield, and in fourth grade they take canoe field trips and make models of the Connecticut River valley in north central Massa-chusetts.

All of these approaches aspire to helping children build a sense of commitment to place and community—to inspire kids to care deeply and to want to make a difference. Robert Yaro, a regional planner in New York City, said, "Stewardship springs from connectedness" (Driscoll 1992). This is the prize teachers need to keep their eyes on. And in the introduction to the latest version of Mitchell's *Young Geographers*, Sam Brian of Bank Street College of Education said,

> What then is our peril as we approach the last decade of the 20th century? Finishing last on an international standardized test of geographical locations? Slipping to second best in economic pro-duction or standard of living? Perhaps the real peril is that, through ignorance of our environment and our relation to it, our human geography, we will so squander and befoul its resources, physical and human, that we will inflict real suffering upon our-selves for generations to come.

"Love it or lose it," summarizes David Orr (1995). This is our chal-lenge. We need to engage children in a developmentally appropriate map-making and social studies curriculum to make them advocates for preservation. Developing a sense of place is one piece of the puzzle in re-making our schools with a focus on sustainability. After an ecological liter-acy initiative in Springfield, Vermont, one seventh grader said,

> I didn't really think that throwing away a can of Raid was that bad until we did the project. When I was little, I didn't really think about it. I just thought my water comes from the faucet and it's clean and it's perfect. Unless, it comes from near the landfill. Then it's just like spraying a can of Raid in your mouth.

This kind of realization increases the likelihood that this student will take an active role in town politics and work to protect the health of the water resources. Keep this context in mind as you read this book. Map-making is a means to an end. Though we may sometimes get caught up in the details of teaching scale, perspective, accuracy, locational relationship, and other cartographic challenges, we must never lose sight of the goal of using maps as clothespins—a tool for hitching children's lives to their places.

2 | THE GEOGRAPHY OF CHILDHOOD
A Developmental Portrait

DEVELOPMENT AND BIOLOGY

First steps by around twelve months, first words by around eighteen months, no more diapers by around three years old. Baby teeth start falling out at about age six and stop falling out at about age twelve. Wisdom teeth make their appearance close to age eighteen. We are all familiar with these developmental markers of childhood. They are indicators of a preordained biological process that unfolds in each child. Environmental factors such as nutrition, parental care, and psychological stress influence these processes, but basically the genes are in the driver's seat.

Many developmental psychologists believe that biological markers correlate with changes in children's brain development and their cognitive abilities. The change in teeth around age six seems to correlate with the onset of representational drawing and an ability to start to understand written words as symbols of speech. The onset of puberty around age thirteen seems to correlate with the maturation of abstract thought. Young adolescents start to question their parents' views on politics and religion; deductive logic starts to appeal to them.

As educators, we try to tie curriculum to biology. Health and sex education is timed to correlate with the onset of sexual maturation at the end of elementary school. Gesell readiness tests use drawing skills as one indicator of children's developmental readiness for reading. Waldorf educators sometimes look for the change in teeth in six-year-olds to determine their readiness for first grade.

But regrettably, in many cases there's a mismatch between curriculum and biology. Reading gets pushed onto unready four- and five-year-olds. Children's inclination for kinesthetic learning from ages six through nine is undercut when children have to sit in their seats all day. And world maps are foisted onto first graders who barely have a sense of their own neighborhood.

Neurophysiologists and learning theorists are starting to be able to describe the unique developmental predispositions for different kinds of learning. Animal behaviorists call these *critical periods*, biological windows when certain kinds of learning are programmed to occur. Maria Montessori called them *sensitive periods*, suggesting that the child was most sensitive to certain kinds of learning at different stages. Howard Gardner currently discusses seven different intelligences and suggests that the biological timetable of each intelligence is different.

For instance, formal music instruction often doesn't start until middle school, but it appears we have been missing the boat. Musical intelligence, Gardner and others suggest, matures early, starting around age four. Kindergarten teachers have always capitalized on their intuitive understanding that music speaks to children. Orff and Suzuki music educators know that five-year-olds can learn to play instruments and grasp music theory when it's presented in the right fashion. But for the most part, we put an inordinate emphasis on reading for young children and music gets neglected. Similarly, children seem to have an earlier predisposition to math and a later predisposition to reading. In most schools, however, we treat math and reading the same and emphasize instruction equally in kindergarten and the early primary grades. Perhaps it makes more sense to put more focus on math and less on reading for first graders.

I believe there is also a sensitive period for helping children bond with the natural world. Between ages five and seven, children start to move away from home and parents and explore the natural world. From ages seven to eleven, children are predisposed to merging with nature and making geographic sense of the world around them. From ages eleven to thirteen, children's geographic skills mature, and they start to move into a stage of social consciousness. In conjunction with this development, children go through a clear cognitive progression in their ability to make maps and understand them. Most school programs and curricula are woefully insensitive to these biological patterns. At the stage when children need lots of opportunities to be outside developing kinesthetic skills and engaging in construction-based activities, we keep them locked up in classrooms. Developmentally appropriate geography and sense of place activities are sometimes completely missing from the elementary curriculum, or they are often conducted in an inaccessible fashion with children. Mapmaking, like piano playing, is a lost art in too many elementary classrooms.

In this chapter I will paint a portrait of children's relationship with the natural world and social community from ages five through twelve. We will look at a series of children's maps produced at different ages, create a model of stages of development, and then extract implications for curriculum. This illustrated conceptual model will provide the foundation upon which the curriculum work of later chapters will stand.

THE ECOLOGY OF IMAGINATION IN CHILDHOOD

When my family and I lived in Costa Rica, we often went on explorations along Quebrada Honda, a rainforesty stream near our home. It was a perfect stream for my seven-year-old son, Eli, and my nine-year-old daughter, Tara—mostly ankle- or knee-deep, with a gently inclined streambed. The stream provided a perfect balance of challenge and safety for children this age. On our third outing, Eli discovered that we could slide down a ten-feet-long, double plunge waterfall on our bottoms. On each following visit we stopped to swoop down and then continued up and around the bends. The kids gamboled ahead, being pulled up the stream by that same invisible magnet that drew me into the woods behind Kevin's house. Dipping in new pools, scooping up thumbnail size frogs in their palms, watching out for snakes. Rock hopping to the other slide, squishing in the mud, examining coatimundi tracks in the silt deposits. One of the last times we went, three white-faced monkeys looped off the hillside to dangle in the treetops. My children are never so happy as this, when they are one with the stream and alert to the discovery around the next bend.

These experiences seem to fulfill a biological impulse for children this age. In her provocative article titled "The Ecology of Imagination in Childhood," Edith Cobb (1959) states,

> The study of the child in nature, culture and society reveals that there is a special period, the little understood, prepubertal, halcyon, middle age of childhood, approximately from five or six to eleven or twelve, between the strivings of animal infancy and the storms of adolescence—when the natural world is experienced in some highly evocative way, producing in the child a sense of some profound continuity with natural processes.

Stream-walking experiences provide an opportunity to merge with the natural world, to feel a "profound continuity with natural processes." The desire to be immersed in nature is a universal inclination at this age. From city to farm, from coast to coast, from Africa to Canada, when social and environmental forces are favorable, children seek to enter the garden. Ornithologist Robert Arbib (1971) recalls the allure of the wide world as he was growing up on Long Island in the 1930s:

> I was nine years old that June afternoon so many years ago when we first discovered the Lord's Woods and the world was unspoiled and filled with mysteries. My first two-wheeler, dark red and fast, had come with my birthday in March, and ever since that glorious day my world had been expanding. Only yesterday I have ventured beyond the edge of my universe, out where Westwood Road ceased to be paved and wound into the endless green unknown of the forest. . . .

We will know it all, Carl and I. We will explore and conquer this America of ours, we will make this our private paradise. To know it and, by knowing, own it, and then go forth beyond our woodland bounds, answering the urgent beckonings of field and farm and road and stream, the distant marsh horizon . . . and the row of trees beyond the last ones we can see.

Just as Leif Eriksson, Christopher Columbus, and Alice in Wonderland felt the desire to expand the boundaries of the known world, elementary-aged children want to find out what's beyond their yard, beyond the end of the road. The challenge for parents and teachers is to understand the scope of the child's universe at different stages. By recognizing the *umwelt*, the life space at each stage, we can encourage appropriate map-making projects that allow children to "know it, and by knowing, own it, and then go forth beyond"—curriculum as wilderness guide.

NEIGHBORHOOD MAPS

Over the past fifteen years I have collected children's maps of neighborhoods in an attempt to enter into their worlds. When I am conducting more intensive research, I ask children to take me on field trips to show me the places on their maps. These trips give me access to the stories and adventures that shape their play lives and allow me to check the map against the actual landscape.

My instructions to the children are open-ended and fairly simple, modified only slightly for younger children.

> I am working on a project about children's maps, and I'd like some help from you. Today, I'd like you to draw a map of your neighborhood. By neighborhood, I mean the area around your house where you spend most of your time and where you play. The only thing you have to include on your map is your own house. Beyond that, it's up to you to show me the places that are special or important to you. It's fine to show other houses, but be sure to include your special places. Your map can include everywhere you are allowed to travel by yourself or with friends, but if you want to show a smaller area, that's fine. Work on your own map and please don't talk with others while you are working.

If younger children appear puzzled by the notion of a "map," I say,

> A map is like a picture of where things are or how things are arranged. If you feel that it's too hard to draw a map, draw a picture of your house and all the special places around your house where you like to play by yourself or with friends.

Regardless of how young the children are, I always first ask for a map because I am interested in the developmental emergence of the "map concept." If children say, "You mean you want a helicopter view or a bird's view?" my response is,

> There are many different ways to draw a map. Any way you choose will be fine. Just try to figure out a way to show me your favorite places.

I provide the children with 15-by-22-inch paper, pencils, erasers and an assortment of crayons. I don't allow the children to use rulers, to encourage more naturalistic, freehand maps. Younger children tend to finish sooner than older children. When the children finish, I ask them to tell me individually about the places on their maps. At the end of each interview, I ask each child to select his or her "favorite place" in the neighborhood. In some of the interviews, the children discuss places that they have not included on the maps. I ask these children to create an extension of their maps, adding on another sheet of paper in the appropriate direction. This often leads me to some of the children's most interesting places.

I have conducted map interviews and field trips with children in Vermont and New Hampshire in the United States; in Devon, England; on Carriacou and Provodenciales (small islands in the Caribbean); and in Monteverde, Costa Rica. Colleagues of mine have conducted similar interviews with children in New York and Boston. Reviewing and analyzing these maps, I have observed consistent patterns of development that appear to be somewhat independent of environment and culture. In other words, there are some underlying patterns that are the same whether the

No Place Like Home
(Rebecca, 5 years old)

Scope: House and yard
Perspective: Pictorial
Attributes: Child's house central and large
              People included
              Sun and rainbows present
              Lots of colors

Out and About
(Matthew, 7 years old)

Scope: Immediate neighbors
Perspective: Slightly elevated (Low oblique)
Attributes: Two or three houses
              Multiple baselines
              Roads appear
              Trees, paths, bushes

FIGURE 2–1   *The evolution of children's neighborhood maps from ages five through eleven. As children mature they move from pictorial maps of their homes to aerial maps of their communities.*

landscape is tropical and parched or temperate and moist. And black children in the Caribbean map their worlds in ways similar to white children in Devon. There are some cultural and educational variations, but there are enough similarities to presume that the unfolding cognitive solutions for making maps are biologically predetermined, much as language development is predetermined. Understanding these biological predispositions can give us a firm foundation for curriculum.

The maps in Figure 2–1 are re-creations of actual maps I have collected from children and show different stages of development. I find it useful to focus on two different aspects of the maps—the scope and the perspective. By *scope* I mean the size and range of the child's world, the life space. The core question is, How big is the territory covered by the map and what are the salient features illustrated? By *perspective* I mean the angle from which the child draws the map. The core question is, What vantage point does the child choose from which to look at his or her surroundings? I have chosen to focus on the maps of children who live in single-family dwellings to show the emergent pattern, but maps from children who reside in apartment buildings are similar in scope and perspective.

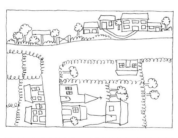

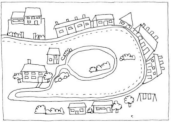

It Takes a Village
(Heather and Vivian, 9 years old)

Scope: Neighborhood/community
Perspective: 45° Elevated
             (High oblique)
Attributes: Houses pictorial
             Roads provide structure
             Forts and hideouts are common
             Legends often used

Up, Up, and Away
(Travis, 11 years old)

Scope: Nearby towns/region
Perspective: Aerial
Attributes: Houses disappear
             Scale becomes accurate
             Symbols replace pictures
             Water courses connect

FIGURE 2–1   *Continued*

Along with the emergence of representational drawing at around age five comes the children's ability to draw simple maps. Prior to this age and extending into these years, children produce maps that often look like the scribbles of early childhood. There's a scribble patch for the house, another for the yard, and a swirl of roads, but not much that looks like a map.

When children begin to produce recognizable picture maps at around age five, it seems there's no place like home. Even though I ask children to map their neighborhoods, the children's own homes dominate their maps. The house is often located smack-dab in the center of the paper and frequently takes up more than one-third of the space on the page. Regardless of whether the house is located far from other houses or amidst a sea of houses, only the child's house is illustrated. In the map, the child typically illustrates favorite shrubs, a climbing tree, the front steps, and places where games are played. Fences delineate the boundaries of the yard, and the natural scope of the child's world is within no more than a hundred meters of the home. Children want to show their bedrooms and draw smoke emerging from the chimney to symbolize the active life inside the house. (Smoke emerges from the chimneys of houses even for children who don't have fireplaces or chimneys.) Parents, brothers and sisters, and pets populate these maps; rainbows and suns fill the sky.

At this stage, the house illustrated often bears little resemblance in overall form to the child's home. The features included are often correct, but the house illustration is schematic—the old standby square with a triangle on top with three smaller rectangles to show windows and a door. Whether the child lives in a duplex, a trailer, a split-level ranch, or a gabled Victorian, most houses will be shown using this generic house schema.

These maps look more like pictures than maps. The early cartographer chooses a frontal perspective for his or her map. There is usually only one baseline in these early maps and not much of a sense of depth. Young children rarely illustrate the back of their houses and have a hard time showing things like a deck in the back of the house or a doghouse directly outside the back door. If these features are important for the child, they will be illustrated above the house, or sticking out on the side of the house.

The feeling of these early maps is of a safe, contained world. Joseph Chilton Pearce (1992) describes children as being firmly ensconced in the mother/family matrix until age four. From ages four to seven the child begins to make independent forays out into the world beyond the yard. At around age six the child wants to walk through the patch of woods to the neighbor's house or down the road to the store. The known world for these children is the house and yard, and they are just beginning their voyages of discovery. Their first maps portray this known world and only gradually start to document the discoveries beyond the edge.

Forts, dens, and secret places start showing up in children's maps at this age. Their appearance is symbolic, I think, of the child's move away from the mother/family matrix and into what Pearce describes as the earth matrix—the natural and physical world beyond the family domain (Pearce 1992). At this age, children's special places are usually underneath the porch or in between a shrub and the side of the house. Later on, the forts of nine- and ten-year-olds move out in back of the garage, in the strip of woods between neighborhoods or behind the billboards. As special places become more important, the house recedes in significance.

The maps of seven- and eight-year-olds show that the scope of the children's world has moved beyond their own yards. Some children just show immediate neighbors, while others expand into the whole neighborhood. Pathways begin to tie together parts of the neighborhood. A child will explain, "You cut through this passage in the hedge, go down behind Nicola's garage, and then slip under the fence into Samantha's yard. Watch out for Mr. Baker's mean dog!" Roads appear now as well but they are truncated and disconnected—just pieces of roads rather than a network. Landscape features that facilitate adventure play a more prominent role; culverts, fallen trees, abandoned buildings, swings, slides, streams, and ponds crop up on these maps.

A sense of depth starts to characterize the maps of seven- and eight-year-olds. A distinction between foreground and background emerges, often by the inclusion of multiple baselines. And because of a change in perspective, the map actually starts to look like a map. The point of view from which the child observes the surroundings is still frontal but it is slightly elevated. It is as if the child is looking at the house and surroundings from a second-story window. Important trees in the backyard are invisible until they stick up above the roofline; the walkway to the door is wider closer to the viewer and narrower closer to the house. And the house itself is somewhat smaller and is moved to the side of the paper instead of occupying center stage.

These maps give the feeling that the child is often out and about in the neighborhood. At this age, my children and their friends started pinballing back and forth to each other's houses. They figured out different ways to go, sometimes met at the frog pond, and sometimes converged on the unused horse barn for a secret game. Their range was perhaps one quarter of a mile and the limits of travel were agreed upon by children and parents alike. The seven- and eight-year-old is constructing a new sphere of influence spreading out from the family home to the neighborhood.

It Takes a Village to Raise a Child: Ages Nine Through Ten

Nine- and ten-year-olds are the explorers of childhood. Their maps run off the sides of the pages and I often find I need to provide them with extra sheets of paper to show me the places they are describing beyond the edge.

They explain, "Here's Billy's field, and beyond that there's the auto junk yard, and back beyond that there's the little stream we like to catch frogs in." In contrast to the self-contained world of the home or the neighborhood illustrated by younger children, these maps suggest a mercurial enthusiasm for exploring outward in all directions. When she was nine years old my daughter began to go on long rambles by herself or with friends. They'd fill their daypacks with water and cookies and set out for the loop trail down to the gorge or for the mile-and-a-half walk to the farm on the back road with the spectacular view of Mount Monadnock. To paraphrase Tom Rush, nine- and ten-year-olds get the urge for going.

Annie Dillard (1987) captures the spirit of this urge in *An American Childhood*, her autobiography of growing up in Pittsburgh in the 1950s. At about nine years old, she recollects,

> Walking was my project before reading. The text I read was the town; the book I made up was a map. First I had walked across one of our side yards to the blackened alley with its buried dime. Now I walked to piano lessons, four long blocks north of school and three zigzag blocks into an Irish neighborhood near Thomas Boulevard.
>
> I pushed at my map's edges. Alone at night I added newly memorized streets and blocks to old streets and blocks, and imagined connecting them on foot. . . . On darkening evenings I came home exultant, secretive, often from some exotic leafy curb a mile beyond what I had known at lunch, where I had peered up at the street sign, hugging the cold pole, and fixed the intersection in my mind. What joy, what relief, eased me as I pushed open the heavy front door!—joy and relief because, from the trackless waste, I had located home, family and the dinner table once again.

Annie Dillard needed a city to satisfy her geographic appetite. For nine- and ten-year-olds growing up in rural or suburban settings, a village will do. The scope of the child's world may now measure up to a mile or more. One of the principal causes of the radical expansion that happens around age nine is the bicycle. Children have gained both the physical competence and sense of responsibility to handle the safety challenges of bikes on roads with traffic. This skill, combined with the geographic urge, takes them out of the neighborhood and into the town and broader community.

In these maps, the child's home is tiny and is often indistinguishable from the dozens of other houses illustrated. In rural areas, large chunks of the landscape are identified and delineated—big patches of the map are labeled *woods* or *big development* or *fields*. Pathways connect friend's houses or lead to secret places in the community, and networks of roadways now provide the skeletal backdrop for the map. Children also start to show an incipient understanding of the structure of the watershed at this age, so streams and rivers crop up and are sometimes connected. If the community is safe enough so that parents allow children to explore freely, the urban and rural landscape dominates the children's maps.

Real perspective has emerged for many children by this age. Many maps are drawn as if the cartographer is up on top of a high hill or a very tall building looking down at the neighborhood or community. This panoramic view is from about a forty-five-degree angle, much higher up and farther away than the ten-degree angle view of the seven- and eight-year-olds. Vivian's map (see Figure 2–1) is a perfect panoramic view of a child's cul-de-sac neighborhood. Houses on the far side of the neighborhood are drawn to show their fronts while houses closer to the viewer are drawn to show their backs. Notice that the mapmaker is a bit puzzled about how to deal with houses as they go around the corner. This is a view Vivian has never seen of her neighborhood, so she constructs it out of her everyday experience of the houses and her understanding of how things look from up high. The child's house is insignificant on this map, but her special place—in the hedge by the school—is illustrated in a larger than life fashion.

Less sophisticated cartographers at this stage solve the problem of wanting greater scope and more of an aerial view by combining frontal and aerial views. In Heather's map, the roads, hedges, and pastures are all accurately portrayed from an aerial perspective, but the faces of the houses are all preserved and illustrated as if from a frontal perspective. It's interesting to notice that when there are buildings facing each other on both sides of a road, children sometimes solve the representation problem with the fold-down technique. Children imagine how the fronts of the buildings appear and then they fold the image down backward away from the road. Thus buildings on opposite sides of the road are folded away from each other.

Both of these maps attempt to coherently demonstrate how things are distributed in space without sacrificing the personality or faces of the buildings and landscape features. Tourist maps use this same solution. Panoramic view maps are easier for newcomers to use because they preserve the visual experience of being in an area. They also capitalize on the fact that, because of poor geographic education, many adults do not understand large-scale aerial view maps very well. The map sophistication of the average American equals that of a ten-year-old, and panoramic view maps make most people more engaged and more comfortable.

It is no wonder that "Explorers" is often a social studies curriculum theme in the fourth or fifth grade. The maps of nine- and ten-year-olds suggest an expansive, pushing-the-outside-of-the-envelope feeling. Children are simultaneously immersed in the landscape and are developing a sense of cognitive distance and perspective. Personally, the maps of children this age appeal to me deeply because they are poignantly balanced between the pictorial narrowness of childhood and the cool abstraction of adolescence. These maps express a joy of living in the landscape, a connection with community that is unique to what some consider the golden age of childhood.

Up, Up, and Away: Ages Eleven Through Twelve

My research with children led me to collecting maps from more than one hundred five- to fourteen-year-old children from the villages of Harvey Vale and Belmont on the island of Carriacou, Grenada. Though politically part of Grenada, Carriacou is somewhat isolated, undeveloped, and only thinly touristed. There are few maps on the island and not much of a local culture of graphicacy or visual literacy. Of all the maps I collected, only one was drawn from an aerial perspective. This was done by an eleven-year-old boy who had moved to the island from Brooklyn a year before. Many of the older boys drew very sophisticated panoramic view maps. Llewelyn's map, for instance, showed all the good hunting and gathering spots around the village of Belmont—iguana hunting in the hills, sea urchin collecting along the rocky coast. It was drawn from a point out at sea and up in the air, an impossible vantage point, yet the awkward shape of each house, the location of the electrical lines and TV antennas, and the prominent trees were all perfectly located. What this view lacked in portraying abstract spatial configuration it made up for in its provocative detail.

In England and the United States, the pure aerial view emerges in eleven- and twelve-year-olds' maps, and the scope begins to stretch beyond the local community. Travis's map stretches well beyond his home village of Ogwell, England, to include parts of Denbury, large natural areas, and the center of the large market town of Newton Abbott (see Figure 2–1). While doing research in this part of England, I acquired the most recent, detailed government maps of the area. But since the large new housing developments had been built in the previous few years, none of my maps were current enough to allow me to locate the homes of the children in my study. Travis's map, due to its comprehensive accuracy and scope, turned out to be the map I used to navigate to children's homes and to follow our field trip travels. Travis also indicated a significant development shift when I asked him about his favorite places. From ages eight through eleven, most children describe secret places in their neighborhoods—forts in small copses of trees, hollowed out hedges. Travis had constructed many dens in the past few years down along the River Dart. But Travis's favorite place was now "the shops in Newton Abbott." He liked the cultural stimulation and social context of being downtown. Similarly, on Carriacou, the girls up to age eleven talked about their play shops as special places, but the twelve- to fourteen-year-old girls liked the Mermaid tavern or the history museum in the main town of Hillsborough, almost five miles from their village. By this age, children's interest naturally matures beyond the here and now to the long ago and faraway (Mitchell 1991).

The aerial view, prematurely imposed on children starting in kindergarten and first grade, finally flourishes now. It is as if the cartographer is in a hot air balloon looking down pensively on the expanding countryside. Because of the angle of the view and the height, the particularities and faces of the neighborhood fade away. Trees are indicated in the architectural plan style. Travis's map hardly even shows any houses. Just as the individual distinction of the house faded away in the previous stage, the child's neighborhood now merges into the broader countryside. The built and natural infrastructures of roads, pathways, and rivers give the map its form. The trappings of formal map expression are also much more common at this age. Certainly, children imitate what they have seen in maps used in the curriculum. But their use of symbols, keys, scales, and grid reference systems also comes out of their conceptual understanding of how these components work and why they are useful.

By eleven or twelve, children have achieved the cognitive sophistication necessary to be able to deal with abstract information. And not only can they make real maps, but they also can start to read and use maps to navigate in the landscape and to make inferences about geographical patterns. Of course, eleven- and twelve-year-olds just being introduced to the idea of making maps for the first time will have to go through some of the earlier stages before being able to make pure maps. Also, though the aerial view is accessible for some children at this age, many may linger with a preference for panoramic views or for mixed frontal and aerial views in their own maps. In a follow-up study of the British children, I found that some of them didn't take to the aerial perspective until age thirteen or fourteen. Thus it is important to note that the sequence I have described is relatively dependable, but some children will move through these stages must faster than others. And some children will only make it part of the way.

A NOTE ON COGNITIVE DEVELOPMENT

The progression of children's mapmaking skills is a microcosm of cognitive development in elementary school. At five and six, children are still immersed in early childhood and their world is small, contained, and dominated by sensory perceptions. The right hemispheric mode of spatial and visual perception dominates, and feelings and pictures are main forces in the organization of the child's world. The houses, trees, and animals are faces in the landscape that carry a certain emotional valence and their "look" needs to be preserved along with the relationship between the child and the aspects of his or her surroundings. By eleven or twelve, the child has gained perspective, both literally and figuratively. The child has gained the ability to illustrate a view from a chosen perspective—to stand back, separate from the scene, and observe it. While the younger child is bound up in the lack of differentiation between subject and object, the older child can take an objective look at the subject of the landscape. The evolution of children's maps can tell us how to approach geography and environmental education curriculum, and it can give us insight into the cognitive changes that characterize the challenge of moving from kindergarten to sixth grade. Forcing the first or second grader to get in the driver's seat and steer is asking for trouble.

IMPLICATIONS FOR CURRICULUM

The point of this close examination of children's maps is to inform classroom practice, to bring curriculum in tune with biology. Right now, much of our curricular practice is discordant. The practice of premature abstraction creates inharmonious music that causes many students to stop listening. Rather, we want to give children simple instruments and accessible instruction so they can make beautiful musical compositions that start simply and gradually gain in complexity. To guide this training, I suggest some basic principles of the small-world approach to mapmaking that will underlie the hands-on suggestions for teachers.

Make Big Maps of Small Places

Currently, we do just the opposite. In that little book of continents that my first grader brought home, Australia was about four inches across. Atlases are filled with small maps of big places. The biggest map kids usually see is the map of the United States, the retractable one hung from hooks on the edge of the blackboard. This is the right size for a map, but the area depicted should be the school playground, or the walk from the school to the animal shelter. First maps for children should be big, bigger than 8$\frac{1}{2}$-by-11-inch paper—as big as the desktop, as big as the tabletop. In fact, first maps should probably not be maps; they should be three-dimensional models.

Remember that Model Making Precedes Mapmaking

In his research with children, Roger Hart (1979) discovered that when children were given three-dimensional materials such as blocks, cut paper, small trees, and a toy car, they made far more accurate maps of their neighborhoods than when they drew them two-dimensionally. Especially in the primary grades, and sometimes in the intermediate grades, maps should be built and sculptured as much as drawn.

Honor the Expanding Horizons Progression

As children's maps increase in scope, so should the maps in the curriculum. Introduce maps of the desktop and sandbox in first grade, maps of the school and playground in second grade, maps of the city block around the school in third grade, and so on. It's true that children can understand maps of greater sophistication than they can make, but you should always maintain a developmentally appropriate mapmaking strand in the curriculum.

Use Pictorial and Panoramic-View Maps

Take a look at the map of the "100 Aker Wood" from *Winnie-the-Pooh* (Milne 1926). The perspective view is exactly like that used by eight- and nine-year-olds. In fact, look at the maps from *Treasure Island* (Stevenson 1997), *My Father's Dragon* (Gannett 1948), and *The Hobbit* (Tolkien 1966). They are all panoramic-view maps with pictorial enhancement. Authors and illustrators use this kind of map because they intuitively understand that these maps make sense to children. Yet we rarely use these kinds of maps as part of the curriculum. Encourage children to draw these kinds of maps and—here's the hard part—get used to mapping this way yourself.

Walk, Don't Run

We all have a tendency to want to skip over the necessary, incremental steps. Yet when children don't need to count on their fingers anymore, they will stop on their own accord. When it becomes cumbersome for them, they incline toward efficiency. Therefore, it is OK to linger at the early stages. For example, instead of submitting to the curricular charge to cover Vermont geography in fourth grade, focus more appropriately on the geography of Brattleboro. Field trips are much more affordable when you focus on what is local. Provide children with many substantive experiences of making maps of visible, accessible places. These will then serve as metaphoric bridges to understanding those smaller maps of bigger places.

3 | THERE'S NO PLACE LIKE HOME AND SCHOOL
Ages Five Through Six

"There are two things that six-year-olds are good at when they come to first grade," said the educational psychology professor in my teacher training program. "They can talk and they can move around. And the first two things we tell them are: Be quiet and sit in your seats." Right away, we undermine children's strengths and ask them to start learning in unfamiliar ways. Instead, we should begin with curriculum that capitalizes on children's ability to walk and talk. This is the key to effective mapmaking projects with young children. Children learn the landscape through walking in it and talking about it. We can extend this naturalistic learning by challenging them to represent their knowing with appropriate media. In the same vein, first mapmaking projects should focus on the narrow scope of the immediate world that the child lives in. The home and school are rich enough environments to serve as the content for children's maps.

These beginning concepts are important whether you are working with first graders or with fifth graders who have never done any mapmaking before. Older children with little mapmaking experience will also benefit from the introductory experiences. They won't need to linger in them for long, but they will need to rough in the working concepts to help eliminate gaps in their skills and understanding. Try to resist the temptation to assume that their previous teachers "covered all the material." Instead, begin with map projects that allow you to diagnose the sophistication of children's abilities.

There are many good spatial relationship activities going on in most kindergarten and first-grade classrooms. Block play will lead to the creation of landscapes with roads, intersections, cities, and hills. Board games are often maps of fantasy landscapes that children easily project themselves into. Hide-and-seek encourages children to explore the topography of the school yard, and other games create landscapes with safe zones, boundaries, and places to avoid. All of these activities provide the kinesthetic foundation for spatial understanding that will get formalized in the mapmaking curriculum. No comprehensive early primary classroom is complete without them. In this book, however, my focus will be on the activities that encourage representing the surrounding community and getting to know the local animal and human inhabitants.

Before starting, you might want to have children draw the kind of neighborhood maps I have described in chapter 2 as a type of diagnostic tool. Similar to the Gesell Draw a Person task used to assess developmental readiness for first grade, I have found neighborhood maps to be an excellent way to assess children's cognitive skills and find out a little about their homes and families. Comparing these with the maps in this book will give you a sense of the children's conceptual abilities, and it will help you decide where to start in the following continuum of activities.

MAKING A THREE-DIMENSIONAL CLASSROOM MODEL

Here's one good way to start. The developmental precursors of this activity are dollhouses and sandbox roads. As I stated earlier, it's much easier for children to relate to three-dimensional models than two-dimensional images on paper. Models look more like the things they represent, and first maps need to bear a lot of resemblance to the thing itself. Additionally, when you place something in the wrong place, it's easy to pick it up and move it. But when you draw something in the wrong place, the hassle of having to erase it slows down the process and takes motivation away from mapmaking. It's also messier.

Find a place in your classroom where all the children can sit in a circle. Before you assemble the children, build a simple model of your classroom using wood blocks, Cuisenaire rods, pattern blocks, or whatever other small manipulable materials you have available. Don't skimp on size. Make your model at least two to three feet long depending on the size of the blocks you're using. The bigger the better. If the model is large, you will be able to indicate small furnishings on your map while staying roughly in scale. Be sure to orient the model to the classroom so that the wall with the door to your classroom is parallel to the wall with the door in your model (see Figure 3–1).

Focus on getting the outline of your room correct, including the location of the doors. Windows are a bit problematic because they don't occupy any floor space, so you might want to leave them out in the beginning. Include your desks, tables, counters, and maybe an art easel if appropriate. But don't overdo it. Leave out some things for the children to add in later. The idea is to include enough of the elements of the classroom so that it's recognizable, but not so much that there's nothing else to add. Now hide pennies somewhere in the classroom and use small stickers, such as tiny gold stars, to indicate where you've hidden the pennies. Choose locations that you can accurately indicate on the model. It's hard to indicate that something is inside a desk, but you can put the penny under a book on a table if you can find a tiny object to signify a book on the tiny table in your model.

FIGURE 3–1
*Playing hide a penny in a
Costa Rican classroom
using a three dimen-
sional classroom model
made with Cuisenaire
rods.*

Have the children sit in a circle, show them the model, and ask them if they recognize this place. If you have first graders or older children, there's a good chance that one of the children will get it. Explain that a model is a copy of a real thing, only much smaller, and ask the children to describe examples of other models that they have seen. Point to parts of the model, and have individual children stand up and walk over to the object in the classroom that you are indicating in the model. Start with solitary objects that are easily recognizable, such as the door or the teacher's desk. Then explain that you have hidden pennies in the classroom and that the little gold stars show where they are hidden. Have one child at a time go to find a penny. Be prepared to have some children who are not able to find the object and encourage other children to provide some clues when you think it's necessary. Ask children to explain how they knew where to look.

Ask the children to close their eyes while you rehide the pennies. Then return to the model, relocate the stars, and choose more children. After you've demonstrated appropriate hiding places for a while, let the children start doing the hiding and the placing of the stars. Keep your eyes open to assess the accuracy of the star placement. Toward the end of the activity, you can elaborate the model a bit. Ask the children to identify some things in the classroom that aren't indicated on the model. After they suggest an item, find an object to represent it and place it on the model. Then ask the children if they think you have placed it correctly. Place some of the representative objects incorrectly so that they can correct you. Then choose a child to add something to the model and not tell what it is. Ask the rest of the children to figure out which object in the classroom the child is representing.

I know of one classroom in which the teacher kept her classroom model up for at least a week. The children continually added things to it, played in it, and eventually took it down and rebuilt it. Children will also enjoy playing this game again as a class and among themselves. Ask the children if they have ideas about how to change the game. Having the children stand above the model to sketch it or asking them to trace around objects in the model are good ways to introduce the idea of the aerial view. However, this is a perfectly good activity without committing paper to pencil at all. Many of the initial concepts of mapmaking, such as orientation, scale, direction, similar shapes, signs and symbols, and grid reference systems, are introduced in this activity in a playful fashion.

SEARCHING FOR A PLACE

One of the clues to effective mapmaking is finding the right context. Children are motivated to make maps when there is a real use for them, such as helping other people locate things. The next two activities involve making maps that take other children on journeys.

Another principle that underlies these activities is the idea of starting where children are at. Much current thinking in cognitive science and curriculum theory focuses on the notion of eliciting children's conceptions and misconceptions about a topic before they begin to study it. For instance, if teachers want their classes to study evaporation, they should begin by eliciting examples of evaporation from the students and getting them to articulate what they think they know about evaporation. Teachers might be surprised that my nine-year-old thinks that "when water evaporates it gets absorbed into your skin." Ignoring this misconception will leave it intact in the child's mind, and new learning will get layered on top of it without changing the underlying conceptions (Gardner 1993b). With mapmaking, children need to have fun and to attempt to use maps made by their peers. These opportunities will motivate them to figure out how to make maps that work.

Both of these activities will only work if you have the chance to do them twice—once to illustrate effective strategies and problems and a second time to provide an opportunity to solve the problems. Since children are actually drawing maps in these activities, they are much more difficult than the prior activity and may be beyond the capacities of some first graders.

Finding Your Desk

If you have a classroom with desks set up in rows, here's a good way to introduce the use of drawn maps. Using paper and a pen or pencil, draw a prepared outline map of the classroom that shows the door, some desks, and some features (see Figure 3–2). I suggest a perspective rather than an aerial view of the desks. This takes longer to make, but it is more accessible to the kids. Discuss with students the objects shown on the map and how you have chosen to represent some of the objects.

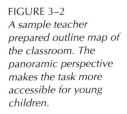

FIGURE 3–2
A sample teacher prepared outline map of the classroom. The panoramic perspective makes the task more accessible for young children.

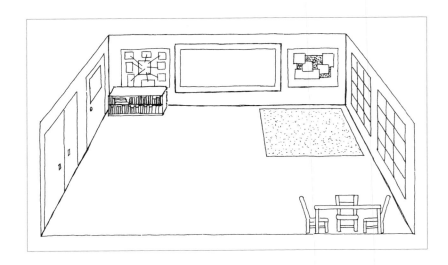

Then have each child complete a map of the classroom and mark his or her desk with an X on the map. Make sure the children do not put their names on their maps. Collect all of the maps and then distribute them randomly to all of the children, making sure that no one gets their own. The challenge is for the children to read the map and sit down in the desk marked with the X. This can become an amusing form of musical chairs because you will find children trying to sit in the same chairs. Once everyone is settled, the children hold up their map so the mapmakers can identify themselves and indicate whether the map readers are at the right desks. Children will comment on problems they had in interpreting the maps: "It's hard to tell the front of the room and the back of the room" or "Selina said this is the blackboard, but it looks like a table to me." This second comment illustrates the frequent problem that children have as they try to pictorially show vertical objects such as blackboards, posters, windows, and displays.

Examining these maps with the children will help you to focus everyone on some of the core problems. Questions that will emerge are well identified in *Mapping: ESS Teacher's Guide* (Barth 1985):

> When you look down on things, what do they look like?
> Do they look the same as when you look from the side?
> What would this table look like if you looked at it from the ceiling?
> What would a tree or the school building look like from the sky?
> How can you tell how to hold this map?
> Which is the left side of the room? Which is the right side?
> Does it matter which way you are facing?

Though it is valuable to have the conversation about how things look from above, don't suggest that this is the only right way to make a map. Pictorial and panoramic view maps are fine, and a combination of map views in one map is often an effective way to make a good map. You might want to share the illustrations in Figure 3–3 with your students. These illustrations show how a house and yard would look to a bird in three different positions. This progression corresponds to the natural progression of perspective as illustrated in children's maps in Chapter 2. It is valuable to validate the pictorial and panoramic view perspectives and even encourage children to draw maps in this way. Jumping to the aerial perspective too soon undercuts the development of their drawing and visualization skills.

Hide a Penny

If the children are successful with the "Searching for a Place" activity and seem ready to draw their own maps, then try this challenge. This time everyone gets a penny to hide. Choose a limited area—the classroom, part of the hallway, or some part of the playground—and specify the boundaries for the children. Every child gets about two minutes to hide a penny and then about ten minutes to draw a map to show the hiding place of that penny. The map can include any kind of information—pictures, words, arrows, number of steps—that the child wants to include. The children should draw a circle with the date of the penny at the hiding spot on the map. The date will help let the finders know whether they have found the right penny or not (see Figure 3–4; note that this student did not include a penny date).

FIGURE 3–3
Pictorial, panoramic, and aerial views of a house and garden. A map can be drawn from these three different perspectives. Children and teachers should be encouraged to use the pictorial and panoramic views in the early elementary grades.

Aerial view

Panoramic view

Pictorial view

It doesn't matter too much if children see some of their friends hide their pennies because they don't know which penny they are going to look for. When all of the maps are completed, collect them and then pass them out randomly to the students. Ask everyone to check to make sure they don't have their own map and then follow the map and search for the penny. Some of the children will find them easily; some will get completely lost. Have the mapmaker lead the lost child to the hiding place, explaining the map as they go. Once all of the pennies are found, bring the children together and ask them to talk about what helped them find the pennies and what made it hard. You will get comments like these:

What Helped

I recognized the closet because Myra drew the shoes and lunch boxes.
Mark wrote down the number of steps from the door to his penny.
Shelley's map was really neat. I could tell where everything was.
I could tell it was in the bathroom because Jonathan wrote Boys on the door.

What Made It Hard

I could tell it was in the hallway but there was no way to figure out how far down it was.

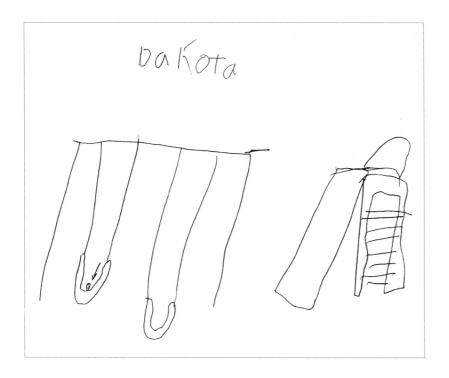

FIGURE 3–4
A first grader's pictorial Hide-a-Penny map. The hiding place was on the seat of a playground swing.

The map showed it was on the table, but it was actually under the table.

There was too much missing from the map. I couldn't find the teacher's desk.

Eladio's map made the shelf much bigger than the table, but it's really smaller, so I couldn't figure out what it was.

After this conversation, ask the students to help you create some guidelines for how to draw good maps. For first graders, some guidelines might be:

Labels help

Showing a picture of how something looks is good

On the map, big things should be big and little things should be little

You should only show things on the map that are always there (for example, Danielle isn't always standing by her locker)

If you can, show the right number of things

There are many good aspects of this activity. The children get to make and use their own maps, the maps serve a purpose, and the children start to understand what makes a useful map. Optimally, you should provide another opportunity to do this activity a day or two later. Review the guidelines and encourage children to try to focus on one or two of these in their maps. As the ESS teacher's guide states,

Practically every problem dealt with in this unit will come up in some form right away. . . . Many of the problems that the children raise have to do with distortion and changing points of view. They have difficulty drawing three-dimensional figures, and easily confuse side views with overhead views and combine the two.

Other problems involve describing the location of something by its orientation and its position in relation to other things. One thing is certain: most of the students will not solve these problems right away. As the unit progresses, they will have chances to confront the same questions again and again.

Rome wasn't built in a day, and children may need many attempts to make sense of these simple maps. Maps of items on a tabletop or of the block structures may be useful for some children. In the next chapter, there are more sophisticated versions of making classroom maps and maps of other small places.

MAPPING THE ROUTE FROM HOME TO SCHOOL

A good first step away from school is to draw maps of the route from home to school. These maps can be good from a diagnostic perspective, like the neighborhood maps, allowing you to assess the different strengths and weaknesses of students in your class. This challenge also gives children an opportunity to look closely at the daily experience of going to and from school. The transition from home to school can be both exciting and difficult for kindergartners and first graders. Making a map and talking about it gives them a venue for sharing their experience, hearing other children's daily experience, and getting a sense of the community that they are all a part of.

One of the early map skills appropriate for children this age is sequence—in other words, helping children represent the order of places along a line of travel. Children need to be able to represent sequence in one dimension before they can coordinate the location of places in two dimensions. This activity is constructed to help children focus on what things they pass on the way to school and in what order.

First, create some worksheets that ask children to draw the following pictures.

1. Draw a picture of your house and a picture of the school.
2. Draw a picture of the car or bus you ride in to school each day. If you walk to school, draw a picture of yourself walking. You can include your friends or siblings if you walk with them.
3. Draw a picture of two important places you notice on the way to school. (These could include bridges, hills, stores, significant trees, beautiful views, or places where your friends or relatives live.)
4. Draw a picture of two dangerous or scary places along the route to school. (These might be railroad crossings, street crossings with lots of traffic, places where there are people you don't like, or steep hills that are slippery in winter.)

Once the children have completed all of the pictures, have them cut out each of the drawings and construct a collage of their route on a large sheet of art paper (see Figure 3–5). (Note that this is not a collage map, but was drawn all at once by the student.) Ask students to follow these directions:

1. Glue your picture of the house and the school on opposite ends of your piece of paper.
2. Connect your home and school by drawing the roads and paths that you ride or walk along each day. Try to remember corners where you make big turns.
3. Arrange your important and scary places along the roads between home and school. After you've described them to your teacher, glue them in the right places.
4. Place the picture of you riding or walking to school somewhere along the route to school.

When the children are done, have them describe their maps to each other in groups of three. Encourage each child to talk about the important and scary places on the map. You might want to come up with a color coding system to differentiate between fun places and scary places on children's maps. Also be prepared for surprises. When I did this activity with a class of seven- and eight-year-olds in Devon, England, one of the scary places that emerged was the village toilets. These public facilities located down a short lane off the main street seemed innocuous to me, but all of the children in the village agreed that they were a creepy place.

If you'd like to take this activity further, you can try to create a teacher-drawn composite map of the things children see on the way to school. This assumes that you have some knowledge of where children live and how they travel to school. On the board or on a large sheet of paper, draw a picture of the school in the center. Then have some of the children describe their maps while you draw the sequence of places from home to school.

A more child-centered alternative would be to have the children create this larger map with teacher scaffolding. Create a bulletin board with the school in the center and identify an approximate location for each of the children's homes by lightly writing their names on the bulletin board. You will likely have to sacrifice accuracy for the purposes of providing enough space for each child to draw his or her route. Then, have a few children at a time translate their routes from their personal maps to the bulletin board. It is likely that there will be some overlap of features on children's maps, so

FIGURE 3–5   *A first grader's home-to-school map. This map was drawn rather than constructed using the collage technique described in the text.*

have children from the same neighborhoods or along the same routes work together to draw the big bridge or fire station that they share on their maps.

It is easy to see that this activity blends into the neighborhood stage of mapmaking. Cultivate the realizations that happen when children say, "Hey, I drive over that bridge too!" The shared experiences cause a shift from a feeling of "This is my place" to "This is our place." This is one of the foundation stones of stewardship.

HUNTING FOR TREASURE

In School Yards

The big picture

A picture-based treasure hunt is another wonderful way to introduce children to map use. This challenge engages the whole class and allows you to make a map of the school yard that is mostly pictorial. You will begin by creating a large map of the place where the hunt is going to take place (instructions for each step follow). The playground or some part of the school grounds is a good place for beginners. The map will show the locations of puzzle pieces you will create and hide. You will need to ensure that there is one puzzle piece for each child or team of two students. The children will search for the pieces, take turns finding them, and then put them together to reveal a drawing of the spot where the treasure is hidden. The group will run to this place and gleefully find the treasure.

Making the map

To help you draw your treasure map, take a look at the map I drew for my daughter's seventh birthday (Figure 3–6). To draw this map, I imagined that I was up on top of a high hill or a tower, looking down at an angle at my house and barn and backyard. I drew this so it would look like the map of the Hundred Acre Wood at the beginning of *Winnie-the-Pooh* (Milne 1926). The oblique view will allow you to see how things are arranged around the school yard and still see how things really look. One of the problems is that you can't show everything. For instance, I can't really show the patio or the bulkhead entrance into the basement on the back side of my house. This also means that you won't be able to hide any puzzle pieces in these places.

I could also have chosen to look down on my house from the other side, as if I were up above the garden looking down at the back of my house. You have to choose your angle based on the places where you want the treasure hunt to happen.

Remember these things when you're drawing the map:

- Choose a big piece of paper so you can fit a big space on it. Make the pictures of objects and locations big enough so that they are recognizable, but small enough so that they're not all on top of each other.

- Do not try to show everything in your hunt area. When there's too much on a map, mapmakers say there's too much *noise*—a different way of saying there's too much clutter.
- Use little bits of color to help identify things. I showed our red front door because it's the only door on the house that's red. That way, everyone knew right away where the door was. I made the pumpkins in the garden orange and the blueberries on the bushes in my neighbor's yard blue.
- Do not worry about it being exactly right. Your pictures just need to give the kids the idea that it's by the front door of the school, near the slide, and so on. Your map picture of the slide doesn't have to look exactly like the slide because there's probably only one or two slides on your playground and it's OK to search at both of them.
- Identify the hiding places as accurately as possible on the map. For my map, I showed these places with little gold stars that I bought in sheets at an office supply store. You can also draw little stars.
- Consider covering your map with clear contact paper or putting it on a backing of contact paper to make it sturdier. Since you will probably wind up using the map for other purposes, it's worth the extra effort.

Making the puzzle and hiding the pieces

Choose a final hiding place for your treasure that you feel comfortable drawing a big picture of. It's best if it's something you can open up or lift or unveil. For my daughter's hunt, I chose our mailbox and drew a picture on

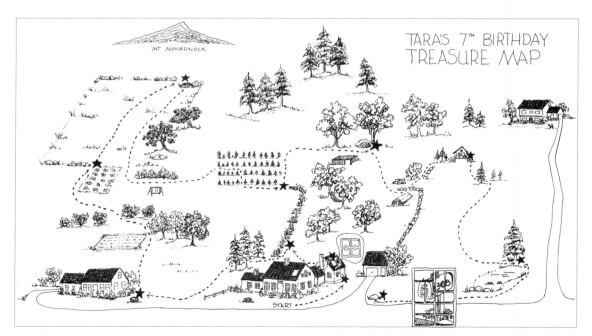

FIGURE 3–6 *A panoramic view map for Tara's 7th birthday treasure hunt. This kind of panoramic view map should be used for treasure hunts and map reading activities with kindergartners and first and second graders.*

a twelve-by-sixteen-inch piece of sturdy paper. Then I glued the picture to a thin piece of cardboard. I cut the cardboard into pieces to create the puzzle. I simply cut mine into rectangles so I could easily cover them with clear contact paper in case it rained or I wanted to use them again. You can also make odd shapes.

I hid the pieces of the puzzle so that they were slightly hard—*not too hard*—to find once the kids got to each place on the map. This gives everyone in the group time to get there and look before that piece of the puzzle gets found.

I hid my puzzle pieces:

> In the old well in front of our house
> In a blueberry bush in my neighbor's yard
> In a raspberry bush in my neighbor's yard
> In a corner tree in the patch of Christmas trees in our neighbor's yard (these are *great* people to have as neighbors)
> Underneath a pumpkin in our garden
> Stuck between two logs in our woodpile
> Just sticking out from behind a board inside our play house
> In the spokes in my son's bicycle
> Underneath a crate in the back of our truck
> Nudged into the ground next to the big rock at the frog pond

Hint: Because some pieces of the puzzle may tell more about the final hiding place than others, and because you don't want everyone to figure it out until they find all the pieces, I suggest hiding any *giveaway pieces* (the pieces of the puzzle that by themselves make it easy to figure out the final location) at the place you think the kids will go to last.

Conducting the hunt

Imagine that the children have just gotten back from lunch. It's a beautiful day and they are excited about the special activity you have hinted at. You have wrapped the map like a gift and you have hidden it someplace in the classroom where you know they will find it. Here's what you do next.

1. When a child finds the map and runs up to you, gather all the children on the rug and contain their excitement while the finder unwraps it. Don't be surprised when someone exclaims, "It's a treasure map!"

2. Let each child see the map and get oriented. Help kids identify the school and the places around the playground that they should know. Say, "I think there are clues or puzzle pieces hidden at each one of these places. Where should we go first?"

3. Make sure that everyone stays together and that you all go from one hiding place to the next as a group. (This is the hard part for the teacher—half the class will want to run.) Because you're collecting pieces of a puzzle rather than going from clue to clue, it doesn't matter what order you go in. Together, decide where

you're going to look for the first puzzle piece. Once a child has found it, look at the map again and decide where to go for the next one, and so on.

4. Let everyone know that each student, or each pair of students, gets to find one piece. Once students have found a piece, they need to hold themselves back at the remaining hiding places. They can still look, but they can't reveal where the piece is. Another alternative is to put names on the map indicating who gets to find the clue hidden at each spot. If there are more puzzle pieces than kids, let everyone find one first and then start over again.

5. Let the children carry the puzzle pieces they have found. Help them understand how important it is to hold tightly to their piece: "If we lose any of the pieces, we might not be able to find the treasure."

6. Once they've found all the pieces, settle everyone down and spread all the pieces out on the ground or a large stone. It's best if you can find someplace firm and flat. Tell them that when they figure out where the treasure is, you'll all go there, and you'll all open the treasure together.

7. Let them arrange the puzzle pieces. Give them help only if they're getting frustrated. Eventually someone will get it and shout where it is. At this point, it's hard to stop the mad stampede. As they run off, yell after them, "Be sure to wait so we can open the treasure all together." This way the kids at the back won't miss out on the fun.

8. Choose one child to lift up the rock, open the door, or whatever lets you into the treasure. Make sure that you've got a piece of the treasure for everyone. It's not really important for the treasure to be valuable or spectacular because the hunt itself is appealing to the children.

9. Afterward, talk to the students about the hunt. Ask them, "Which was the hardest puzzle piece to find? Did you have ideas about where it was hidden? How did the map tell you where to look? Which was the most exciting part?" They'll have fun reliving their adventure.

Simpler Alternatives

Let's face it. Some children will have trouble with this simple kind of map, and the idea of making a map gives some teachers the shivers. Maybe the treasure hunt idea sounds like the cat's meow to you, except for the mapmaking part. Or perhaps you have a group of young kindergartners who will have trouble with any kind of map.

If your school is in an urban area, you may have a related problem. If you don't have a school yard and you can't do your hunt in a nearby park or playground, you may be limited to using the sidewalks around the city blocks where the school is located.

The solution to all of these problems is to use exclusively pictorial clues or a pictorial map to structure the hunt. A straightforward picture map is easier to draw and easier to interpret for very young students. Here are two alternatives.

One option is to use the sidewalk itself as the main path to follow. The whole hunt can take place by going around the block—you never have to cross the street. For instance, let's imagine that you start the treasure hunt at the entrance to your building. You start with a picture of something on your block: a window sign, a special kind of fence pattern, or a fire hydrant. Everyone searches until they come to that place, where they find another picture of the next feature to find. One clue leads to the next until the treasure is found. One problem might be that people who aren't involved in your hunt might find your clues. That means you either have to put them out right before you start, or you have to hide them especially well so that passersby will not see them.

Another option is to do a map hunt with a picture puzzle. For a map of a city block, you just want to show the sidewalk and the fronts of the buildings, so a panoramic view map isn't necessary. Instead you want a pictorial map (see Figure 3–7).

To create a good picture map of your block you need a looooong piece of paper so you can show all the apartments, stores, and other features. If you're going to go all the way around your block, do the map on four pieces of paper. Hide picture puzzle pieces along the way. Once the group has them all, they can go to the treasure and find it.

For city treasure hunts, I sometimes like to have the puzzle pieces or the treasure hidden inside a store. Many store owners love being asked to participate in treasure hunts, and this is a great way to have constructive school and community interactions. Don't be afraid to ask store owners to keep the treasure behind the counter and tell them to give it to the children only if they come in and politely ask if the owner knows anything about a treasure.

FIGURE 3–7 *A pictorial city street map for treasure hunts and map reading activities in urban areas.*

CLASSROOM PORTRAIT:
TREASURE MAPS AND PIRATE HATS

Amy Carter, Kindergarten and First-Grade Teacher
The Harrisville School, Harrisville, New Hampshire

When Amy Carter noticed that Cameron, one of her first graders, was fascinated with maps, she tried to create projects for him that would encourage his imagination and extend his interest. One morning, after Cameron constructed a village in the block area, she asked him to draw a map of it with Bobby, another first grader, during math time that afternoon. Though they completed a map, they were not engaged in the task and their work was sloppy. Amy went back to the drawing board.

A week later at the class's morning meeting, Amy excitedly shared a discovery with her students: "Yesterday afternoon I was up in the school attic looking for some material to make puppets, and I found this tucked away in a corner." She held out a slightly dusty piece of rolled paper that was held together with a gold seal and tied with a piece of red ribbon.

"It's a treasure map. Let's open it!" said Bobby. "No, we shouldn't open it. It's someone else's," cautioned Jennifer. Amy suggested that since it looked as if it had been in the attic for a while, she thought it was OK to open it to see what was inside. Amy had created a nonreader's pictorial treasure hunt. The children excitedly followed the map, collected the clues, and put together the puzzle that pictured an old claw foot bathtub now covered with a table top. No one had ever looked inside. When they lifted the top they found Hershey's kisses and a fancy pencil for each child.

Though Amy thought of this as a onetime special event, the children thought differently. A couple of days later Jeffrey and Shawna created a treasure hunt that led to a buried box on the playground. Children created maps during drawing time, wrapped them up with ribbons, and exchanged them regularly. Nicole, who had a secret place in the stone wall next to the playground, made a map of her hideout. One child who couldn't draw or write well made a treasure hunt by tiptoeing around the classroom and copying labels off the furniture. His simple clues read: "GO TO SINK, GO TO DESK." He was thrilled when everyone actually did it. On the other end of the spectrum, Jennifer and Bobby worked on a treasure hunt for a month before sharing it with the rest of the children. They collected costume jewelry and found a miniature pirate chest and buried it in the sandbox. Their map was expertly drawn and each puzzle piece was a part of a rainbow. Once the class fit the pieces together they had to find the end of the rainbow.

By the end of the fall, Amy found that the vocabularies of maps and treasures had started to percolate throughout the shared language of the class. They were creating their own microculture with their own dialect. They talked about "hunting" or "searching" for answers in reading and math; they knew that if they "followed the clues" they could figure something out. When they came up against a classroom problem, Amy would sometimes draw a map of the problem so the children could work on it. Children's drawn maps showed a fascination with traps, connectors, hideouts, forts, lava, dripping blood, dark caves, wild horses, and swinging nets. Whereas the mapping of the block structure hadn't excited Cameron and Bobby, the treasure hunt idea really captured the class's imagination.

After Christmas, the enthusiasm persisted. Some of the children worked for two weeks to create a village in the block area. Once it was complete, they created a treasure hunt that took place completely inside the block village. The messages were taped on the undersides of blocks in different parts of the village.

When the class listened to a tape of *Peter Pan*, Amy initiated a story map project. Together, she and the kids created a map of Never Never Land. It showed Peter Pan's hideout, Captain Cook's ship, Skull Rock, and other places. After this activity the children showed an interest in pirates and jewels, so Amy did an improvisational storytelling activity over a number of days. Amy started a story and then children took turns adding to it. When the children got stuck, Amy jumped in and helped them figure out where to go with the story. Once the story was complete, each child chose one of the scenes to illustrate, a couple of children made a map of the story, and they bound it all together to create an original, authentic classroom book.

The map motif lasted throughout the year. The older children got interested in board games and created their own boards for games such as Traps & Chutes and Fire & Flame. The objective of the latter was to follow the pathways to the castle so you could marry the princess. During the winter, following the pathways of animal tracks fascinated the children, and reading the stories in the tracks led to writing animal stories. The marble labyrinth maze became a favorite classroom toy.

I think the key to Amy's success with children this age was in her realization of the link between maps, treasure hunts, and children's imagination. Her initial attempt at curriculum innovation was too dry and cognitive for children this age. But the secret treasure map appealed to the children's visual orientation and to their interest in adventure. Amy capitalized on this shared interest by using the images and metaphors of maps and treasure hunts to characterize and describe her whole curriculum.

Amy drew out the stories that were inside the children. When not making maps, the children were involved in the normal work of math and reading, but Amy made it all feel as if they were on a pathway in the midst of exploring new terrain. And she helped bring them out by encouraging them to draw pictures and maps of these real and imagined places. Then, like walking through the wardrobe into Narnia, Amy and the children entered into the imaginary worlds they had drawn and created. With children of this age, maps should be vehicles for connecting both to nearby places and the vivid imaginary worlds of early childhood.

4 | OUT AND ABOUT EXPLORATIONS
Ages Seven Through Eight

SEEK AND YE SHALL FIND

During our four-month sojourn in Costa Rica, my children attended the Blue Valley School in Escazu and I collaborated on some projects with teachers at the school. Though the name of the school makes the setting sound romantically rural, it is anything but. Escazu is a corner of metropolitan San José, and the school is shoehorned between small shopping centers, condominiums, and the busiest street in town. Loud diesel trucks belch exhaust and buses rumble by the entrance to the school. The traffic makes field trips look like a nightmare. Street vendors wander through the traffic selling papayas and avocados. Armed security guards lurk in front of most commercial establishments. At first look, it's not the kind of neighborhood you'd want children to bond with.

But down the street, and abutting the back side of the school's cramped playground, is the Vivero Exotica, a plant nursery and greenhouse. Walk inside and you're in one of those rain forest exhibits at the botanical gardens. Trees rooted in the ground grow up through the ceiling and sprawl out through windows. Red gravel walkways wander around a small pond and past a cage of toucans and pheasants. Visit at the right time and enjoy a delicate indoor rain shower sifting down from the ceiling. The traffic is miles away; your soul feels soothed. Second-grade teacher Heather Barber and I decided to make a model of this special place with her rambunctious seven- and eight-year-olds. It was close enough for repeat visits and it got kids into nature in the middle of the city. Best of all, the nursery offers more than meets the eye. As Heather commented after noticing the subtle red trim on the edge of the palm leaves, "This place is great. The longer I'm here, the more new discoveries I make." This is exactly what we hoped would happen for the children. I describe our experiences in the portrait at the end of this chapter.

In South Brent, Devon, England, where I worked with many groups of children, there were neighborhood studies projects going on in almost every classroom. The kindergartners mapped the arrangement of fruits and vegetables at the greengrocer's shop. The first graders made a map of the town center while the second and third graders were mapping the street from the school to the post office. Meanwhile, the fifth graders were making detailed survey maps of the playground and the man-made frog pond. Though South Brent sits on the edge of misty Dartmoor National Park and is one of the rainiest villages in Devon, the teachers were often out on short, impromptu field trips with the children. This was possible because the school was smack-dab in the middle of this compact little village. Everything was relatively close at hand, though nothing was particularly unusual or striking.

The lesson here is that the neighborhood around any school offers up fine grist for the cartographic mill. Maybe you're thinking, Well, that may be true at those schools, but out here in the boondocks, or down here in the inner city, there's either nothing to see or it's too dangerous. Au contraire—Seek and ye shall find! In downtown Manhattan, certainly a challenging environment even in the 1930s, Lucy Sprague Mitchell focused on the geography of production and distribution in the city. Knowing that young children are fascinated with the jobs that people do, she focused on

> a whole series of water-front trips to docks and boats, to harbor activities such as lighthouses and buoy stations; to railroad transportation activities, such as freight trains and switching yards; to garages and gasoline stations; to storage and distribution centers such as markets, stores, refrigerating plants, and milk plants. . . . Of course, six-year-olds do not get the full significance of what they see. But before seven, most of the children have a vivid sense of the congested island which has to be fed and clothed from outside, which has to make roads over and under the streets in order to carry on its business. (1991)

In urban Schenectady, New York, I had the pleasure of exploring Back Secret with a group of elementary teachers. This twenty-acre parcel of land was hemmed in by housing developments, storm drainage ponds, and railroad tracks. It harbored a few makeshift dwellings for teenage gatherings, jumps for dirt bikes, and gravel pits. But it was also a fascinating labyrinth of trails, birds, small mammals, and interesting flora. In rural Nelson, New Hampshire, where the school is far from the village, I have led children on orienteering adventures in the wildlands behind the school.

I went through all of high school biology getting great grades thinking that the cutaway, three-dimensional model of the flower in the classroom referred to some faraway flower in the tropics, not to the flowers growing outside the classroom window. No relationship was made between the world of learning and the world outside the doors of the school. Don't let this happen to you and your children. Your neighborhood offers up little treasures to discover and great opportunities for learning. Mapping projects based on nearby places can open the doors.

In the previous chapters, I have put the most emphasis on the developmental relationship between age and the scope of maps—the progression from school to neighborhood to community and beyond. This progression provides the structure for Chapters 3 through 6, but there are other considerations as well. As children get older, they have the capacity to move along a continuum of increasingly sophisticated representations. This basic notion was intimated in the guideline at the end of the second chapter that emphasized that models should precede maps. But let's flesh this idea out a bit more here (see Figure 4–1).

As the figure shows, the youngest children (ages five and six) will be most comfortable and capable with models and *tool maps*, Lucy Sprague Mitchell's term for maps that you work and play on top of. (The vinyl map with roads and bridges for use in the block area is a good example of a tool map.) Pictures and murals are also effective map forms at this age. Emphasize locational sequence and concrete representation with children this age. Slightly older children (ages seven and eight) will start to be able to make and read panoramic view maps. They can also do field sketch maps for the purposes of collecting qualitative data for use in making maps back in the classroom. You can introduce the skill of scale at this age level in a qualitative fashion, and children can easily grasp pictorial symbols at this age.

Not until children are nine and ten will they be able to really use the baseline and offset technique to collect quantitative data to create mathematically accurate maps of classrooms and outdoor areas. Children this age like to incorporate keys, symbol systems, and compass roses into their maps. Grid reference systems and spatial distribution make sense to them.

The oldest elementary-age children (ages eleven and twelve) can realistically grapple with the mathematics of scale. Some can learn to use compasses and can orient maps correctly. They can start to use triangulation mapping techniques (described later in Chapter 6), they can collect data for the purposes of making contour maps, and they start to be interested in map projections.

All of these different ways of making maps should be used throughout the elementary years. In other words, there are times when it is valuable and useful for nine- and ten-year-olds to be making models as well as doing more demanding mathematical projects. A good rule is, As you move up the representational scale from model making to creating distant aerial views, move down the map scope scale. Therefore, when you are beginning to teach children the somewhat demanding baseline and offset mapping technique, it is important that you use it in a small area such as the classroom or a one-meter square of the playground. Similarly, when you are asking young children to map something beyond their normal developmental scope, move down on the representational scale. If you're asking nine-year-olds to make maps of a big area such as Vermont rather than of

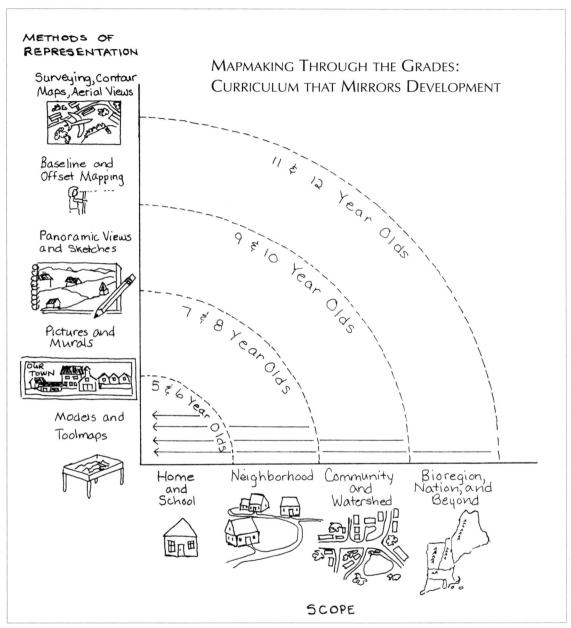

FIGURE 4–1 *Developmentally appropriate approaches to mapmaking throughout the elementary years. In terms of media, a gradual progression from models to panoramic views to aerial views works well. In terms of scope, a gradual progression from the close at hand to the wider world is appropriate.*

their local communities, have them make plasticine topographical maps rather than two-dimensional, aerial view maps.

In the chapters that follow, I will be describing mapmaking projects that move outward in geographical scope and projects that challenge children's representational abilities while focusing on small, nearby places. It's appropriate for seven- and eight-year-olds to be making both looser neighborhood maps and classroom maps with a bit more emphasis on correct location.

As children approach the intermediate grades, it becomes more important to be sensitive to the child's grasp of the structural elements of maps. Perspective, location, direction, legends and symbols, scale, and purpose are the map elements I will focus on here. These map elements are analogous to punctuation and grammar in writing. Therefore, as in the whole language approach, I recommend putting less emphasis on these elements of "punctuation and grammar" in the earlier years while putting more emphasis on the creative production of maps. Children will develop a gradual understanding of each of these elements.

We have already discussed the perspective progression from pictorial to panoramic and aerial view. Similarly, we will look at the child's emerging capacities in each of the skill areas. A child's understanding of scale, for example, is predominantly qualitative until about age eight or nine, and then it becomes more quantitative.

For instance, earlier this week I was doing the "Hide a Penny" activity with second graders. When I asked the children to add something to the model, one girl chose to represent the blackboard. She represented it by stacking three black Cuisenaire rods on top of each other. I asked her to compare the heights of the real door and the real blackboard in the classroom, and she noted that they were the same height. I then asked her to compare the door and the blackboard in the model, and she saw that the door was four rods high and the blackboard was only three. She immediately added another rod to the blackboard. She understood that the blackboard should be "as high as" the door in the model, a qualitative relationship. This understanding of scale was appropriate for her to grasp. Trying to get the children to understand that there was a 1:20 scale ratio between the model blackboard and the real blackboard would have been a challenge. That kind of understanding can wait until fifth grade.

Understanding these stages of development in perspective, scale, legends, and other map elements will help you have realistic expectations of children's mapmaking capacities through the grades.

LEARNING THE FACES OF THE NEIGHBORHOOD

Focusing a project around one street or pathway is a good way to move away from the classroom. In the same spirit as the map of the route from home to school, the idea here is to follow a line of travel and map the elements along the way. This puts the locational emphasis on sequence rather than on spatial distribution. It's like the difference between playing Candyland and Battleship: in Candyland the players move forward and backward along one set pathway; in Battleship the players operate in a landscape of grid references. Simple linear routes of travel can start in the school. Map the route from the classroom to the school office or from the classroom to the water fountain, putting an emphasis on what the wall features are along the way. Or map the flower beds that line the walkway from the parking lot to the front door of the school. In each of these cases, the focus is on the sequence of things observed along the way. Are the daffodils before or after the crocuses? What's located right before the fire alarm box?

Faces of the Neighborhood: Three Examples

The following three examples all took place during the same autumn. Note how all three projects focus on almost the same section of the village very close to the school, yet the character of the representation grows slightly more sophisticated, depending on the children's age.

A tool map with grades K–1

One of the challenges of mapping a street is figuring out how to show buildings on both sides of the street. It's hard for children to go from the three-dimensional experience of seeing the faces of buildings to knowing how to represent them on paper. As with younger children, one solution is to use a model. One of the first-grade teachers I worked with in Devon, England, came up with an insightful solution. She wanted the children to make a map of the main street of the village of South Brent, but she also wanted the map to be a tool map, a place where children could play at being in the village. Her solution was to take photographs of all of the prominent buildings along the street. By taking pictures Thursday afternoon after school, she was ready to use them on the following Monday. She mounted the pictures of the buildings on individual-serving-sized cereal boxes (maple unit blocks are more stable if you have them) and then conducted a group activity of having the children assemble the main street of the town. She scaffolded the activity by locating the buildings at both ends of the street—the parish church and the post office. Then, one by one, she asked children to place buildings in their appropriate locations.

It was easy to slide the buildings one way or the other to create a niche for each addition. The teacher noted disagreements about location, and these questions provided the structure for a short walk downtown later in the week.

The model of the main street stayed up in the block area while the questions were resolved. Together the teacher and the children drew a map of the street, and then the model was disassembled. Pairs of children were challenged over the next week to use this map to reassemble the village. Both the teacher and classmates evaluated the reassembled models by using the map to check the locations of the buildings.

An emergent bulletin board with grades 1–2

British teachers seem to be particularly good at creating what I like to call *emergent bulletin boards*. Instead of using the prefabricated holiday bulletin board or the "Explorers" bulletin board (complete with portraits of Columbus and Balboa and little cutout Spanish vessels), British teachers often make the creation of the bulletin board the main focus for a month-long unit.

In one classroom in the same school in Devon, the seven-year-olds were creating a bulletin board of the two main streets of the village. In a group discussion, the teacher had the children brainstorm a list of the prominent buildings in town. Their objective was to list at least twenty-eight—enough so that each child in the class would be responsible for one of them. On the class field trip to town, each child had to draw a sketch of his or her building. Back in the classroom, each child made a large drawing of the building (as big as twelve by eighteen inches) and then mounted it on a box or a piece of stiff paper with the edges folded back. This way, the buildings had a sense of bulk and dimension and stood out away from the bulletin board.

The teacher's preparation for the project included creating a skeleton of the village on the bulletin board—representing the roads with black construction paper, specifying the location of some landmark buildings, and figuring out how big the buildings should be so that they would fill up the space but not be too big. The teacher asked parent volunteers (when available) to take children out on the field trips. I accompanied one pair of children when they went to make a sketch of one of the portable classrooms. Because there were so many windows in the classroom, I encouraged them to count the windows so they could put the right number in their picture. There were ten windows in all—two rows of five—but when the children completed their picture they showed one very long row of seven on top of one very short row of three. They had held on to the idea of ten but not to the accurate spatial configuration—a good example of the limitation of children's observational skills at this age.

As children finished their buildings, they placed the buildings on the bulletin board after much whole-class discussion. Once all the buildings were completed, the teacher and class discussed what else they needed to make this a real village. Storekeepers, cars, delivery trucks, animals, puddles, rainbows, manhole covers, and gravestones for the church cemetery were all added to complete the picture. As an observer of the process, I enjoyed watching the village center emerge on the bulletin board much the same way a photograph emerges in a darkroom developing pan.

A child-designed model with grades 2–3

In the third classroom, the teacher scaffolding for the seven- and eight-year-olds was much more conceptual than actual. Whereas the teachers for the younger children created the frame within which the children painted, this teacher asked the children both to create the frame and find the materials.

The teacher introduced the activity saying,

We're going to make a model of the area from the school to the post office. How should we do it? We've got to figure out things like where it should go, how long it should be, who's going to be responsible for which parts, and what materials we will use to make the buildings.

She took the children on a few short field trips down to the post office to collect data. Then with her subtle guidance, the children created a detailed model. Responsibilities were divvied up and the model was created on top of a row of student desks. In this small classroom of thirty-four children there was no extra space, so for the four weeks that this project was under way, the children just had to work around it (see Figure 4–2). The school grounds were modeled, including trees lining the playground; shrubs in the yard across the street; the basketball court complete with uprights, backboards, and baskets; and the goldfish pond. The pond was a modeling masterpiece. A combination of aluminum foil on the bottom with another layer of plastic wrap on top perfectly created the illusion of water. Between the two layers were goldfish made of popcorn kernels and lily pads made by cutting pieces of leaves into round circles. The children chose blocks of wood scraps to make the buildings on the street. A renovation site with construction scaffolding was simulated with the use of toothpicks and Popsicle sticks. The teacher encouraged an emphasis on accurate location and on relative size and scale. After looking at the model, the teacher asked, "Is the length of the basketball court the same as the distance from the school to the renovation site?" This sent a pair of kids back out to take some

FIGURE 4–2
One section of a desktop model showing the school basketball court and fish pond. Completed by second and third graders in a British classroom, the whole model showed the village from the school grounds to the post office.

new measurements. The idea of scale and measurement naturally emerged out of the process of building the model.

There are a number of progressions at work in these three examples. The first model focused on just the faces of the buildings, the second focused on faces but made the representation more two-dimensional, and the final example moved toward more of an emphasis on an aerial view, though still with three-dimensional materials. In each successive example, the size of the representation decreased. The tool map for the youngest children was about eight feet long, and the children could work inside of it. The bulletin board map was about six feet long, and the tabletop model was about four feet long. The amount of responsibility and decision making increased as the children became older, with less concrete scaffolding provided by the teacher. Finally, the amount of detail and mathematical accuracy emphasized was also greater at each step. It's possible to facilitate the growth of children's conceptual skills with a deep immersion in the local. As William Blake suggested, the challenge is "to see a world in a grain of sand."

Down a Country Lane: A Sound Map and More

You don't need buildings to make this kind of project work. Mapping a nature trail, a hedgerow, or a country road works just as well, if not better, than mapping a downtown street. As the focal point for a mapping project, a natural area is much more likely to lead a class into natural history, plant identification, science, and environmental education. An urban project lends itself more to history, social studies, and geography. Choose your mapping project depending on the subject area that needs the most elaboration.

Some of the best mapping projects I have encountered came out of the Parish Maps project, a community mapping project initiated by the British environmental group Common Ground in the late 1980s and early 1990s. (See the project profile at the end of Chapter 5.) Though the project focused on making maps of entire communities, I most enjoyed a smaller project conducted by a visiting artist in the small Devon community of Ipplepen. The artist decided appropriately that mapping the whole community was impractical in the one week he had available, so he decided to map a small area adjacent to the school. For the center of his map, he chose a crossroads about three hundred meters from the school where four lanes radiated outward like the spokes of a wheel. With groups of about ten children, he started at the crossroads and mapped each lane out a distance of about two hundred meters. His passion as an artist was to use natural materials from the landscape in the artwork he created. "If we're going to show this lane on our map, then let's collect mud from the lane to use to paint the lane onto the map," he encouraged them. Children quickly grasped the rightness of the concrete connection between the place and the map.

For the map of one lane, he started children off at the crossroads, saying,

I want you to walk down this lane silently and make notes of all the things you hear. Anything and all things. Loud sounds and soft ones. I want you to collect enough sounds so that we can make a sound map of this place. That way, when people look at our map they'll be able to hear in their minds what it would sound like to walk down this lane. But don't just write down boring things like *bird, wind, leaves*. I want you to try to capture what things really sound like. Use your imaginations.

Without a peep the children collected sounds until they reached the end of the lane. Then, for the way back, he said,

Now, I want you to find things around you that we can use to mark or color on the map. For instance, if we're going to paint this little stream on the map, we should collect stream water and mix it with watercolor paints to put it on the map.

On the way back, children experimented with rubbing all kinds of plant materials onto paper, and they collected wood chips, plant materials, and other artifacts for inclusion on the map.

Back in the classroom, the artist orchestrated the creation of a circular map four feet in diameter. As he collected the children's sound observations for different places along the lane, he shaped a sound poem that recreates the auditory experience of walking in this place. Stretched out on the map along the lane, part of the poem reads,

> Echoing
> Deep voice,
> Cows arguing,
> Burping
> Opera
> Pop songs,
> Sore throats.
>
> Water running,
> Singing,
> Whispering,
> Secrets.
>
> Engine,
> Car driving,
> Splashing.
>
> Leaves
> Cracking.
>
> Heartbeat.

Strolling through this poem, you can hear the raspy-voiced cows mooing, the tiny stream sneaking along the side of the lane, and distant car noise. The ending "heartbeat" was the sound the children heard at the completion of the walk. In the silence, they could only hear their own hearts beating (see Figure 4–3).

Shards of the landscape showed up in many places in the map. The woodworker's shop was represented by wood shavings. The public footpath was shown by dipping a child's foot in mud from the path and then printing the footprint on the map. Rubbing cabbage leaves on the map gave a light green hue to the cultivated cabbage fields. A montage of photocopied leaves represented the dense hedges lining some of the lanes. A tiny outline map of Australia indicated that some of the stone mined in the local quarry was exported to Australia. The label from the Exeter milk carton showed that milk from their town was enjoyed by people throughout southwestern England.

Focusing closely on the nearby natural world deepens children's experience and helps them to see the connections between their place and other places. The sound map took an ordinary lane and turned it into a place of poetry and inspiration. One message that this kind of project conveys to children is that their places are special, making them just as important to preserve as national parks and historical sites. Another message is that the town and its residents are connected to the rest of the world through local products—the world in a drop of milk.

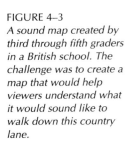

FIGURE 4–3

A sound map created by third through fifth graders in a British school. The challenge was to create a map that would help viewers understand what it would sound like to walk down this country lane.

The mathematics of scale can be introduced with eight-year-olds if the concept is approached in a very concrete fashion. The crux of the problem is recognizing that as the complexity of the ideas increases, the scope of the area mapped should temporarily decrease. Therefore, instead of moving out and about into the neighborhood, we should be like Alice in Wonderland and "get small." This also makes developmental sense because of children's intrigue with small worlds at this age. Dollhouses, model railroads, toy cars and trucks, Lego and Playmobile models, sand castles, and fairy houses are all expressions of children's fascination with being in charge of a small, shapeable world. As one woman once related in a memory of her childhood dollhouse,

> I'm lost in play; time has no meaning. It's my world to change any way I like, to manipulate to whatever plot I want. I'm speaking aloud quietly to myself. I'm there winding down the stairs, going from room to room, arranging things so they are pleasing to me. I'm insulated from the adults upstairs—from the whole world, for that matter. (Sobel 1993)

The core idea is to engage children in the play of small worlds and then extend the play into artistic representations. The curricular principle at work here is the building of bridges between children's play and the academic expectations in the language arts, ecological literacy, and mathematics curricula. My Green's Farms map grew up and out of my exploratory interest. Similarly, children's maps can grow up and out of their creation of play worlds. Or, if the growth isn't natural, at least we can graft curriculum extensions onto the root stock of the play experience.

Drawing Enlargements and Reductions

Before computers and photocopy machines, many children learned the useful skill of "gridding up" and "gridding down" drawings and maps. Even though you can go to the photocopy machine and now get reductions with the push of a button, I strongly encourage teaching children this skill. Not only is it a wonderful trick, but it provides a metaphoric bridge for understanding the whole concept of scale. Once children have the skill in their hands, it is easily transferred into their minds.

As a precursor to the gridding process, you might want to have children do enlargement drawings using an opaque projector. Have small groups of children choose a small line drawing of an animal that they would like to enlarge. Make sure the drawings are simple enough so you won't frustrate children with the task. Photographs are not suitable for this task; the original must be an uncluttered line drawing. Mount large sheets of paper on a smooth wall in the classroom and project each animal drawing on the paper so that the projection is five to eight times larger than the original. The children simply need to trace the projected lines onto the paper to make an enlarged copy of the original drawing. It will be clear that the picture is essentially the same but just blown up or inflated. You might want to find balloons with print on them to provide another model of the transformation.

Gridding up a drawing is the next step. The first time I had to really use this skill was when I was creating exhibits for the Odiorne Point Nature Center on the New Hampshire seacoast. I had read about the size of lobsters during the colonial period and I wanted to convey to visitors how immense lobsters used to be. I gridded up a six-inch drawing into a four-foot monster. It was impressive. Making life-sized animals from a small drawing might be a good way to present this task.

You need to provide the children with two prepared grids. Though the grid size will vary depending on the age of the children, the desk and table size, the paper availability, and the size of the original drawings, try to provide the children with grids that allow for a 1:3 ratio. For instance, you could make one four-by-four grid out of two-centimeter squares, and the other four-by-four grid out of six-centimeter squares. If possible, the smaller grid should be photocopied onto plastic transparencies that can be laid on top of each child's drawing. That way, the child can see the grid and the drawing at the same time. If this is not possible, you will need to have the children re-create the smaller grid directly on the drawings.

Make sure that the animal sits on the bottom line of the grid, using tape to prevent shifts in position. Lightly photocopy the larger grid on paper so each child can draw right on it. The illustration in Figure 4–4 shows the process of taking a small picture of a bunny and enlarging it using the gridding process.

You will most likely need to model this process on the blackboard in your classroom. I find it valuable to analogize the process to dot-to-dot drawing. I try to focus the children on figuring out and marking where the lines of the drawing intersect the lines of the grid:

> Where does the edge of the ear go across this line? Near the middle, or closer to the line on the left? See how the toe just dips below this line and then comes back up.

By creating a set of intersection points and points to aim for, the children can construct their enlargement by connecting the dots to each other. Gridding a drawing down is more challenging for some children, but just as intriguing. Have children do both.

Recognize, of course, that this is not an overnight skill. It will take a while for some children to master the perceptual and fine motor skills necessary to complete a good drawing. Even without refinement, the experience of seeing the drawing grow or shrink will provide a conceptual basis for understanding the scale relationship between the place and the map.

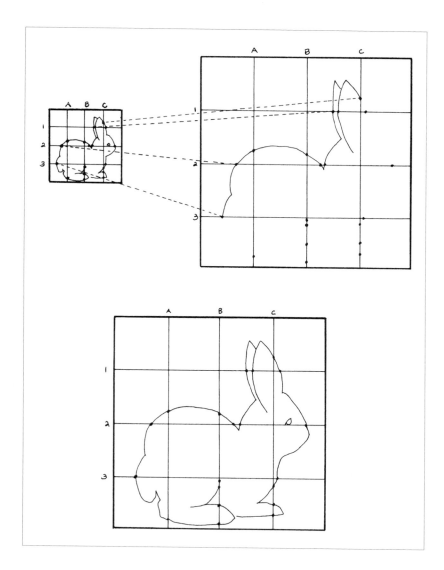

FIGURE 4–4
Gridding up or gridding down a drawing of an animal is good preparation for the technique of mapping small places.

Lego Villages

Any classroom material or game that uses a grid reference system will help establish scale as a mode of thinking. Tic-tac-toe, checkers, chess, and Battleship all use grids. And the reference system for identifying the location of a chess piece on the board is particularly relevant. Legos are particularly valuable from the perspective of making maps. The instructions for assembling Lego models are particularly good for teaching grid interpretation. Lego instructions are completely visual (no words included), and they require children to count circles on a grid vertically and horizontally in order to put a piece in the right place.

The Mapstart series (Catling 1985), British instructional materials for teaching map interpretation, uses Legos ingeniously to create a model village. Built on top of the flat base pieces that come in large Lego kits, the village is complete with a school, a fire station, residences, a gas station, a candy store, a bakery, flower beds, and roads. With an intact village, it is easy to show the children the difference between a panoramic view and an aerial view. It is also relatively easy to create a teacher-made aerial view map and to have children draw maps by looking straight down at the village. Structures are easily placed on the map by having children use graph paper or paper with a grid of dots that represents the raised circles on the Lego baseboards. Buildings can be located by counting the number of dots over and the number of dots up to where the corner of the building is found.

One third-grade teacher in Brattleboro, Vermont, did a particularly nice project with combining community studies and mapmaking. Field trips through the town led to group discussions focused on the question "What kinds of buildings do we have in our town?" Once the list was complete, each child was responsible for creating one of the buildings. Then, within a limited area, the children had to decide where the buildings would be located in their own fantasy village:

> Should the school be next to the fire station? Should the bank be next to the police station? Is it better to have houses close to the supermarket or close to the park? Will all the people want to go to just one church?

After much discussion the buildings were fixed in place, and then small groups were responsible for creating their own maps. Aerial view maps are difficult for eight-year-olds, but when they are drawn from a model that the children can stand above and see all at once, children can manage this task.

Imaginary Worlds: Islands and Microparks

If you want to get beyond the straight edges and plastic of Legos and increase the mapping challenge, have the children construct their own fantasy worlds. Using clay or plasticine, children can create their own personal islands on top of square pieces of plywood or Masonite approximately 50 centimeters in size. Provide guidelines indicating that each island must have

A harbor
A beach
Some cliffs
A pond
Six houses
Some roads connecting the houses
A small forest

Modify this list to taste or have the children suggest some possible additional features. Hills and mountains are fine to include, but they do present a bit of a challenge when it comes to map creation.

If you'd prefer to work outside and engage the children in close natural history observations, challenge them to create microparks. Provide small groups of children (three seems optimal) with six feet of sturdy string, eight miniature flags made with toothpicks and contact paper, and a small person for each child. (Lego people are about the right size; anything too big makes the scale inappropriate.) Provide guidelines indicating that each park should include

> Clear boundaries
> An entrance gate
> A nature trail with stations indicated by flags
>> (Three stations focused on flora and fauna)
>> (Three stations focused on physical challenges)
>> (Two stations focused on special views)
> A bridge constructed of natural materials
> A stream or river
> A picnic area
> A name for their park

After the children have created their parks and trails, you should have them visit each other's parks and have the park creators' play people take visitors' play people on guided nature walks. The nature walk responsibility helps children take ownership of their park.

Making freehand maps

For each of the options, you now have some mapmaking choices. If you want fanciful and artistic maps, have the children make freehand maps of their parks. Require them to label the maps correctly and include all of the specified features. Discuss how you will show standard features such as pathways, hills, houses, streams, and bridges. This is a good opportunity to introduce symbols and to come up with a system that each mapmaker will use. As children create their maps, encourage them to match their maps to actuality: "Your island is about twice as long as it is wide, but on your map you show the length and width to be about the same." Encourage children to draw their maps from either a panoramic or an aerial perspective. One is not necessarily better than the other.

Using grid frames

Another option is to use these imaginary worlds to introduce accurate, mathematical mapping and the use of a grid frame. Be cautious if you're trying to do this with eight-year-olds. The management of a grid reference system is challenging but possible for children this young. By the time children are nine and ten, grid frame mapping makes much more sense to them. So, if you're starting young, make sure the task is simple.

Construct grid frames out of wood or plastic. For wood frames you'll need four pieces of wood that are three-fourths inch thick, two inches wide, and twenty-four inches long. Strapping, thin pieces of wood often used in roofing, works well. Overlap these and nail together at the corners. Plastic frames can be made by using one-half-inch water pipes attached at the corners with 90-degree elbows.

Using strong string, create a grid of evenly spaced squares. You should have four vertical strings and four horizontal strings that create a grid of twenty-five four-inch squares. Label the strings A through D on the bottom and 1 through 4 on the side. Provide children with a prepared paper grid with twenty-five one- or two-inch squares. Label the paper grid to match the grid frame (see Figure 4–5).

To use the grid frames for mapping the imaginary worlds, lay them on top of the islands or microparks just the way you laid the grid transparency on top of the animal drawing. It will be important to place the grid so that one edge of the island or park touches the bottom edge of the frame (see Figure 4–6). This gives you an established point to work from that is easy to find on your paper grid. If the island or park has much topography, you may have to support the corners so that the grid lies relatively flat above your island or park.

Now the task is relatively comparable to gridding down a drawing. Locate the points where the edge of the park or the shore of the island intersect the grid frame lines and mark these on the paper grid. Connect the dots together to create the outline of the park or island. Next, find the points where the nature trail intersects the grid frame lines. Mark these on the map and then connect them. Place by place, you can help the children build up a paper representation of the island or park they have created. Once the bare bones of the place have been drawn, aesthetic enhancements are called for. I am always fascinated by how these places come alive once they are drawn on

FIGURE 4–5
Grid frame construction for use in mapping small places and a worksheet for collecting data when using the grid frame.

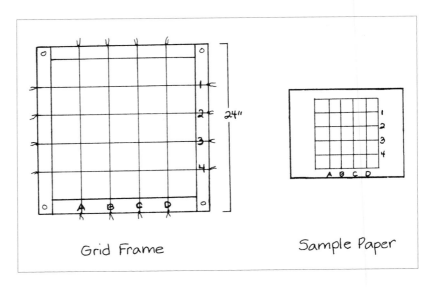

Grid Frame   Sample Paper

paper and artistically embellished with pictorial symbols, blue watercourses, and rolling hills. Finish the map by adding a title, legend, and scale (either 4:1 or 2:1) (see Figure 4–7). These maps can become the basis of an interpretive park brochure that describes the sights along the nature trail and the unique character of this place. Perhaps parents can take self-guided walks through these imaginary worlds when they visit on parent night.

FIGURE 4–6
Mapping an imaginary island landscape using a grid frame.

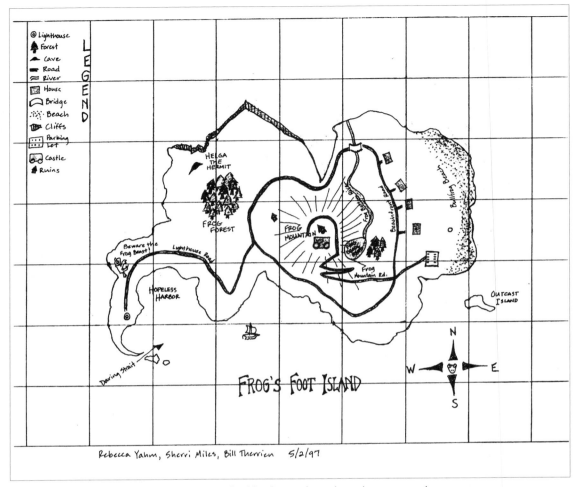

LEGEND

- ⊙ Lighthouse
- ♠ Forest
- ◄ Cave
- ▬ Road
- ≈ River
- ▦ House
- ⌂ Bridge
- ∴ Beach
- ⛰ Cliffs
- ▦ Parking Lot
- 🏰 Castle
- ♣ Ruins

Helga the Hermit

Frog Forest

Beware the Frog Beast!

Lighthouse Road

Hopeless Harbor

Frog Mountain

Bullfrog Beach

Breakfoot Road

Frog Mountain Rd.

Outcast Island

Daring Strait

FROG'S FOOT ISLAND

N
W   E
S

Rebecca Yahm, Sherri Miles, Bill Therrien   5/2/97

FIGURE 4–7   *Finished map of the imaginary island landscape drawn by a classroom teacher.*

CLASSROOM PORTRAIT:
VIVERO EXOTICA—AN URBAN OASIS

Heather Barber and David Sobel, Second Grade Teachers
The Blue Valley School, Escazu, Costa Rica

The Blue Valley School offers an English immersion program for children in the capital region of Costa Rica. The language of instruction is predominantly English, but the language on the playground (and in physical education and social studies) is Spanish. Costa Rican families pay tuition to send their children to the school because they realize how valuable it is for children to know English as a second language. Although the school caters to a fairly well-to-do community, the buildings are small, classroom materials are basic, and the grounds are limited. The enthusiasm and joie de vivre of the students, staff, and faculty are distinctive, however, and a caring atmosphere pervades the school.

By the end of second grade, most of the students can read, write, and converse in English. Their vocabulary is a bit limited (for example, some might have a hard time finding the English words for *wheelbarrow, fence,* and *flowerpot*), but they can get along fairly well. The classes are based on a formal curriculum, which places traditional emphasis on workbooks, desk work, and teacher-based instruction. There is not much cooperative learning, theme-based projects, or student-centered curricula. There is little attempt to develop the interpersonal skills emphasized in "social curriculum" programs in the United States.

With this in mind, I knew that I couldn't jump right in with a challenging, neighborhood exploration curriculum project. The teacher and the children were thrilled with the idea of making a model of the Vivero Exotica, the exotic plants nursery down the street from the school. But I knew I needed a preliminary set of activities to introduce them to map- and model making and to test their abilities to do projects based on small-group cooperation and problem solving. In a didactic follow-the-instructions-and-fill-in-the-blank context, open-ended projects rapidly fall apart.

I started with the classroom model project described in Chapter 3. It worked like a charm. I created the model, hid five colones coins, and indicated in the model where the coins were hidden. The children could read the map and figure out the location fairly easily. Some of them had no problem hiding coins and then indicating in the model where the coins were hidden. I followed up by giving them outline maps of the classroom and requesting that they draw a classroom map and mark their desk with an X. We were moving right along.

Vacation intervened before the next session, and by the time I returned all of the desks had been rearranged. I persevered nonetheless and handed out maps randomly. Then I asked a couple of children to find the location where the desk had previously been. Though I insisted over and over that only two children should be out of their seats, the classroom felt like a pot of popping popcorn. Children yelled instructions, everyone wanted to be searching at once, and the two-dimensional maps were confusing. Chaos reigned. I backpedaled. Everyone back to their seats; no talking!

We abandoned the desk search and went back to model making. I gave each pair of children two sets of Cuisenaire rods and asked them to make a desktop model of the classroom similar to the one I had made. I provided a list of all of the things they had to include. Though a few children had trouble working together, this task saved the day. The well-defined, desk-bound, hands-on, three-dimensional character of the task was just what they could handle. At the end of the class the children looked at each other's models, and I pointed out the unique attributes of some of the models—the accurate representation of all twenty-six cubbies, an ingenious model of the floor fan, the accuracy of representing the legs on the teacher's desk, the importance of showing the window frames. I figured we were ready for a trial field trip on the playground.

I divided the class in half, sent one half to the playground with Heather, and kept the other half in the front yard. We asked each child to hide a coin and then draw a map that would help a classmate find it. The space and time were limited, and hiding places were few. But the task of drawing a map or following someone else's confusing map again proved to be overwhelming. Some of the students were focused and successful but many were clueless. Being outdoors evoked playground behavior, and listening skills plummeted. These children just were not ready to be plunged into two-dimensional mapmaking. We went back to the drawing board to find a more prescriptive, less open-ended mapping task.

I drew an outline map of the school's front yard that included the front gate leading into a small parking lot for about twenty cars, the front steps of the school, and a grassy area with about twelve pines and six stumps. I made a list of the objects for the children to locate and place on their maps and devised pictorial symbols for the trees, stumps, cars, and flagpole. I let the students know that they had to accomplish this task successfully—with no rowdiness—if they wanted to do the Vivero project and then marched them outside to introduce the task. The first five minutes went fine until a dog showed up, bounding from one cluster of children to another. Some children dropped their clipboards, screaming, "He licked me!" "He got fleas all over me!" We removed the dog to the playground, but within a few minutes he was back. Despite the disruption, most children completed the assignment, but the quality of the focus and work was marginal.

Both Heather and I were frustrated with the rowdiness of some of the children but encouraged by the investment of the other half of the class. Since the dog had spawned many of the problems in the last session, we decided to give them one more chance. I pared down the task yet again, requiring fewer things on the maps and making the symbols even simpler. I gathered everyone on the lawn and showed them how some students had gotten the arrangement of the trees pretty well whereas others had them all clumped up next to the building. Together we problem solved how to locate all the trees. A couple of kids demonstrated their thinking, there were glimmers of understanding, and then we tried again. No dogs, rapt attention, a real commitment to the task. We were thrilled, and the children were proud of their work. They liked trying to locate each and every car in the parking lot and figuring out the exact location of the flagpole. They were starting to see the relationship between the map and the place.

Back in the classroom, I asked the children to make another desktop map with Cuisenaire rods, but this time of the front yard. This was the proof of the pudding. Could they use a map they created like an architectural blueprint to help them build a three-dimensional model? I was excited to see children placing their maps next to their emergent models, moving their eyes back and forth from model to map, searching out the flagpole location, calculating the distance from the wall. They were really reading their maps and using them to solve the problem.

Because the children had labored through the painstaking parts, the Vivero model-making project came relatively easily (see Figure 4–8). I created an outline map of the Vivero, which was a difficult task because of the labyrinth of paths, hidden rooms, porches off of porches, and different levels. Then I created a list of things the children had to find during the scavenger hunt. The list made the children see below and above the surface—they had to find the bridge across the pond, tiny purple flowers, an immense potted plant, two trees that grew through the ceiling, a palm tree with spines, the toucan, and more. On the first trip to the Vivero—their first real field trip—a hushed excitement descended over the group.

FIGURE 4–8
A section of the classroom model of the Vivero Exotica constructed by Costa Rican second graders

Their mouths gaped as they saw things they had never seen walking by the Vivero every day on their way to school. Maps completed, we began to collect things we could use to represent the Vivero back in the classroom.

For the next three class meetings we worked hard on the two-by-three-foot model and made a few return visits to the Vivero. We made countless potted plants out of clay and cut paper, figured out how to make trees, made the ponds waterproof, laid down the gravel paths, created a miniature *perro bravo* (the guard dog), and built the flowerpot *casita*. Horacio, who was normally squirmy, worked on the bird cage with birds inside for hours. Giovanna specialized in flower arranging. In between work sessions, students would languish by the model, fiddling with little details, walking through it in their mind's eye. Writing about the project, the children commented, "I discovered a dog and a tree that was two trees that went past the ceiling. I saw at least 10,000 plants. I found the fence between the Vivero and the school playground. I never knew what was over there before." You could see the pieces falling into place for them as they saw the relationship between the model and their daily experiences.

One of the lessons here is that the Vivero Exotica was not built in a day. Community-based mapping projects will only be successful if children have the social skills to do independent work and solve problems in small groups. Students cannot be thrown into a project like this unprepared. There was a progression of mapping skills that we needed to develop before the students could rise to the conceptual challenges in this project. My persistent refusal to move ahead until a majority of the children had the social and conceptual skills was disappointing for the students, but it's eventually what made the project work. Though I was ready to throw in the towel a number of times, my conviction was kept alive by the students' evident excitement when they found me in the classroom and said, "Are we going to the Vivero today!?" In this case, slow and steady won the race. The urban world of these children glimmered with the shadowy, dense, moist greenness of this nearby "jungle" for a few April afternoons.

5 | IT TAKES A VILLAGE TO RAISE A CHILD
Ages Nine Through Ten

CHILDREN IN THE RAIN FOREST

Hillary Clinton's recent book has popularized a saying that my colleagues and I have long been using to describe the relationship between the school and the community: It takes a village to raise a child. It's not just the school's job to educate children; it's the responsibility of the whole community. Schools will do the best job when they are community centers. This means that the community is in the school through frequent parent nights, volunteer programs, guest speakers, and DARE programs. And it means that the school is often out in the community through frequent field trips, apprenticeship programs, school-business partnerships, and community service projects.

This village notion most commonly refers to the interconnected web of social responsibilities and relationships of a healthy community, but I want to give it a geographical spin as well. For the geography, social studies, and environmental education curricula to be healthy in the intermediate grades, projects should move out of the neighborhood and into the broader community, town, and watershed. The maps of children ages nine and ten cover a much wider terrain than the maps of their younger counterparts. One of the causes of this expansion is bicycles. Children now have both the desire to explore and the necessary safety skills. This combination convinces parents that it's all right to let children bicycle by themselves to the town beach, across town to school, or down to the shopping center. The curriculum should travel along with the children.

I visited the unusual village of Monteverde last year. It's a Quaker community located high in the Tilaran mountains of Costa Rica, just underneath the cloud forests of the continental divide. The Quakers started a cheese factory here in the 1950s, which helped to sustain the dairy industry and breathe new commercial life into the local villages of Santa Elena and Monteverde. The community realized that the health of the dairy industry depended on the protection of the cloud forests, so they created a private reserve. There are now four reserves protecting almost one hundred thousand acres, including the Children's Rain Forest, which has been purchased through the fundraising efforts of children all over the world. As a result, Monteverde is one of the premier tourist destinations in Central America and is viewed as a model of sustainable eco-tourism development for the region. Agriculture and low-impact tourism provide a strong economic foundation for the preservation of local culture and the environment in this isolated area.

It's not surprising that an unusual elementary school has emerged in this context. The mission statement for the Centro de Educación Creativa states that the school is:

> committed to academic excellence with environmental education as the foundation of the entire curriculum. [The goal of the school] is to empower future generations of environmentally responsible, bilingual people with the skills and motivation to create change, preserve their natural environment and live in peaceful co-existence with all the people and creatures of the world.

This school is clearly one of the first schools in the hemisphere to put environmental education at the heart of the curriculum rather than at the periphery.

In preparation for a workshop with teachers at this school and other educators in the area, I went for an early morning stroll near the school to find an appropriate destination for a short field trip. The focus for the workshop was on the difference between fostering ecoliteracy and creating ecophobia. My opinion is that ecophobia emerges in children when media, educators, and parents put too early an emphasis on ecological problems. By laying the responsibility for saving the rain forest and protecting endangered species on seven- and eight-year-olds, we alienate rather than connect children with the natural world. Children need to learn the beauty and intricacies of the natural world before they can save it. So, on my morning walk I was looking for a place that would intrigue and engage children. Not far from the school, I hit the jackpot.

The seeds of strangler fig trees establish themselves in the canopy of other rain forest trees and then send roots earthward in search of soil and nutrients. When multiple roots encounter each other, they grow together and slowly create a woven latticework around the host tree. This lattice of roots gradually strangles the original tree by depriving it of water and nutrients. The host tree dies and eventually rots, leaving the strangler fig intact but with a completely hollow core. That morning I found a strangler fig that was slightly tipped over, completely hollow, and easily climbable on the inside to a height of about fifty feet. Being inside it up high was a fulfillment of all my Swiss Family Robinson fantasies. When I brought my own children there, they immediately started to play lost children. When I brought children and teachers from the school, I found out that there was a similar tree down near the market in town.

The next day, guided by the ten-year-old daughter of some friends, we set off to a secret vine-swinging place known by children in the area. On the way, we met our guide's twelve-year-old brother and a friend who showed us two other climbable strangler figs. One of these was more straight up; the other was darker and more mysterious inside. From talking with the children, I realized that their culture in the area harbored a special knowledge of the location and climbability of many of these trees, as well as a knowledge of good swimming holes, great vines to swing on, and other adventurous places.

If I were a fifth-grade teacher at the Centro de Educación Creativa, this is the project I would do. Lots of environmentally conscious families visit Monteverde. The normal activity is a hike through the cloud forest looking for birds and hearing lectures about the intricacies of tropical ecology. We did this and it was great, but how about some activities designed specifically for kids? Perhaps the fifth graders could create a map and guide to the safe tree-climbing and vine-swinging places in the Monteverde community. They'd have to collect data by interviewing children and adults who grew up in Monteverde, visit the sites to determine their suitability, create a map showing the location of all the sites, write an appropriate guide, and design appropriate publicity. Then the fifth graders could be hired as guides by one of the local guide organizations, who would provide transportation and adult supervision. Fifth graders could have a source of income, part of the fee could be donated to the school, and visiting children would enjoy true interactive, bonding activities with the rain forest. There's nothing like being inside a tree to make you feel like a part of nature.

This project may never happen, and the mapping activities you do with your nine- and ten-year-olds might not be this glamorous. But the principles should be the same. Children this age want to explore, put things in order, and have real responsibilities. They want to find out what's around the next bend and make sure that their maps are accurate. This chapter contains descriptions of two very different kinds of projects— exploratory projects that focus on mapping adventures and technical projects that describe very specific techniques for making accurate maps of small areas.

THE EXPLORERS' CLUB

This activity can naturally emerge out of a school yard mapping project or a neighborhood maps project, but you'll have to be willing and able to take lots of short field trips. The core idea is to simulate the experience of living in Europe at the end of the fifteenth century. The continent is well explored and mapped, but out there, in the great beyond, is an unmapped world with vast resources and surprises. What's out there, and can we map it and claim it as our own? Many social studies curricula call for studying the explorers in fifth grade, and this kind of project is a down-home introduction to that kind of unit.

This project can be done either by small groups or as a whole-class activity. If you're focusing on children's neighborhood maps, then they'll have to work individually or in small groups. To explore around the school, you can work as one large team. My focus here will be on projects for the whole elementary classroom.

In the beginning, have children discuss what they know about the known world of the school and the surrounding neighborhood. Create a freehand map on the board of the school grounds and ask the children to draw what lies beyond. This initial mapping project is intended to elicit questions about areas near the school that the children don't know about. With teachers at a public school in Schenectady, New York, we focused on the area named Back Secret, a big patch of city land that abutted the school but that the teachers had very little sense of. Once you've got a completed sketch map of the known world (of your Europe), then you're going to empower the children to go forth and explore the unknown world.

A good piece of literature that parallels this process is *Secret Water* by Arthur Ransome (1984). A father drops his five children off with camping equipment and a sketchy map in the midst of a large British estuary. Their task is to map all of the islands and channels in the estuary by the time he returns to pick them up in five days.

To do this as a class activity, you'll need three different forms of prepared outline maps. First, you'll need a large wall map that shows the school and the known world and sketchy outlines of the unknown world to explore. You need enough definition to create boundaries, but not too much definition so as to take the thrill out of the discovery. The bigger the wall map the better. Four square feet is good, and if you have the space and a way to provide access, five or six square feet is not too big. To create this map you'll need to find a small-scale map of the local area or you'll have to sketch a map of the school and its surroundings. Using an overhead projector or an opaque projector, project the map image on a large piece of paper and trace the school yard and the areas you want to explore onto the basemap. Identify six or seven areas around the school that you want to explore, delineate these with colored markers, and label them. Each small group of children will be working on mapping just one of these sections.

Second, you'll need $8^{1}/_{2}$-by-11-inch versions of the same prepared outline map for the children to use (see Figure 5–1). They will use these maps to collect data in the field.

Third, for the final stage, it makes sense to provide each small group with an enlarged outline map of just the section they are working on. This piece of the map should be just as large as the final wall map. This actual-sized section allows the students to do a rough draft of their map section.

Recruit as many parents and school volunteers as you can, and head out for your expeditions. I recommend two to four expeditions, spaced a few days to a week apart. Initially, the challenge for each group will be to create sketch maps of their areas. Make this process easier by providing children with an inventory of things they should look for. In a suburban neighborhood, the inventory might be:

Man-Made Structures

Roads
Bridges
Houses (especially of people
    you know)
Public buildings (fire station,
    churches)
Stores
Graveyards

Natural Features

Prominent trees
Hills
Streams and ponds
Rocks and cliffs
Fields
Hedges
Flowers
Good hiding places

FIGURE 5–1
A prepared outline map for exploring and mapping Back Secret, an urban woodland in Schnectady, New York. A large prepared outline map was used on the wall and individual prepared outline maps were used by students for collecting field data.

After each group has returned from the first exploration, have them use their sketch maps to start to create maps of their findings in their own area. This is where you need the actual-sized copy of the piece of the wall map that they are going to complete. This intermediary map allows students to plan and revise their section before their results finally go onto the wall. A second or third expedition will serve to flesh out the details of the areas the children are exploring. It may be important for you to meet with each group to help them frame their goals for each expedition or to charge them with things to find out. The third and fourth expeditions can serve as a chance to collect materials from their areas that the children can use to represent some of the places on the map. For example, children may want to represent a sandbank by gluing sand to the map, or a pine forest by selectively cutting some club mosses (princess pines) and attaching them to the map in the right location.

After the final expedition, the teacher should approve the rough draft of the map for each group's section. Once this is done, children can then begin work on the wall map. Encourage children to figure out ways to make their section of the map integrate smoothly with adjacent sections. Complete this project by having children identify all of the things they learned about the school neighborhood as a result of the project.

STREAM EXPLORATIONS

By the time children have reached fourth and fifth grade, an understanding of the local watershed becomes appropriate. On their maps, children this age show an increased inclination toward connectedness and networks; they correctly show road relationships and brooks flowing into streams flowing into rivers. Just this morning my fourth-grade daughter wanted to know if Harrisville Pond and Silver Lake, two of our favorite swimming destinations in town, were connected. She is trying to figure out how it all fits together. Consequently, now is the time to get children out following and mapping streams.

Regrettably, this genuine activity is often transmuted into learning the water cycle. Starting as early as second grade, children do little experiments in jars and soon thereafter draw those diagrams of clouds, condensation, rivers flowing to the ocean, and evaporation back to the clouds. The denatured words have little connection to the real world. Rarely do children step outside, investigate puddles, collect rainwater, make miniature landscapes, or follow streams. Instead, they draw the same diagram, ad nauseam, throughout the elementary years.

The water cycle isn't something to be taught in two weeks; it's appropriately done over the six or eight years of elementary and middle school. The watercourses of the landscape are the circulatory system of the living earth, and we can only learn them by following them, literally and metaphorically. Just as it is valuable to have children trace their own bodies and then map out the circulatory system or the digestive system or the skeleton, it is equally valuable to present children with an outline map of the town or the local region and have them map out the water drainage system. Just as children work from reference books to discover their inner biology, it is appropriate to have children use small-scale, detailed area maps to extract information about the watershed.

These kinds of studies are grounded and enhanced through stream following and mapping experiences. David Millstone, an elementary social studies teacher in Norwich, Vermont, recognized the allure of stream following and the potential curricular value, so he decided to try something unusual. He initiated a class expedition to follow a stream, not knowing where the stream would lead them. In a student-produced newspaper about this expedition, one child wrote:

The Deep Dark Dungeon

"I can't see five feet," I thought to myself. We were walking through a giant culvert following this stream that runs behind the school and through the Nature Area.

"Watch out, dripping water," Mr. Millstone warned us. I finally realized what is beyond the steel grates that you see along the street. I looked up it and saw the grate twenty feet above me. The culvert seemed to be moving. I think we took a turn somewhere.

"The end," someone shouted. . . . I had to walk with my feet widely apart. We got out alive, had snack and continued on our adventure.

Millstone, in an introduction to the newsletter, describes his motives:

We went for the Great Hike Downstream for many reasons. I was curious about the stream myself, and found in conversations with others that no one really seemed to know where the stream went. The trip expanded our recent emphasis on mapping Norwich neighborhoods. The search would challenge the class's map-making skills; similarly, an adventure into the unknown would stimulate children's writing. . . . The experience of following a stream would reinforce a fundamental concept in topographic maps—water flows downhill. The stream flows directly under the new playground, the area which we will be surveying for our own contour map. I wanted children to experience the thrill of posing a question and working directly to find the answer. And not least of all, I thought this trip would be fun.

Like any true adventure, what started out as a simple idea grew more complex as we trudged along. We ended up doing things that I had not anticipated, and going where I had not planned to go. There was valuable learning for both children and adults in dealing with the unexpected.

Each child in the class was responsible for drawing an individual or group map of the journey (see Figure 5–2). These maps show a fine synthesis of exploratory enthusiasm and cartographic skill. They portray the children's excitement at walking through the long culvert under the interstate highway, negotiating their way down a tricky waterfall, and actually figuring out where the stream led. From a cartographic perspective, it is evident that the children were challenged to put the maps in scale, to orient them appropriately, to use symbols, to start to try to represent topography, and to show the relative location of other main landmarks.

Teachers commonly use reflective writing as a way of preserving new experiences, but maps are underutilized in this way. Maps, in fact, are often much better vehicles for representing the experience of moving through the landscape—especially as documents of exploration. Keep in mind the notion of graphicacy—the skill of communicating relationships that cannot be successfully communicated by words or mathematical notation alone. The interpretive map is more effective than words in capturing a streamside expedition.

FIGURE 5–2
Map of stream exploration by four girls after the Great Hike Downstream by David Millstone's class in Norwich, Vermont.

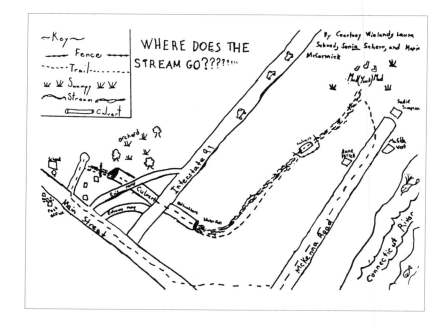

Focused studies of local flora and fauna are enlivened by using grid frames to make detailed maps of small areas. The mapping techniques are the same as described in the previous chapter, but the questions examined are different.

As a graduate student, one of the most compelling field ecology courses I ever took was a course on marine ecology. The main project for this course was to map two tide pools, one that was high in the intertidal zone and one that was low and was only exposed for two hours each tidal cycle. Both tide pools were small, ten to fifteen square feet, but the process of mapping them was fascinating. The mapping process served as a kind of magnifying glass, making us look closer and closer at our Irish moss, periwinkles, and dog whelks. The more we looked, the more we saw. It wasn't until the fourth day that we discovered whelk egg cases and sea cucumbers. They had been there all along, but we hadn't learned to see them. And as we understood the life in our tide pools, we started to gain an appreciation of the ecology of intertidal life all along the Maine coast. The cartographer in Howard Norman's novel *The Northern Lights* (1988) says, "The more exactly I map a place, the more overwhelming what's around it seems." I felt similarly overwhelmed with the beauty and intricacies of coastal ecology once my tide pool maps were complete. The microcosm served as a doorway into the macrocosm.

I adapted this technique when I taught an alpine ecology field course on the flank of Mount Adams in the White Mountains of New Hampshire. I chose a highly wind-exposed ridge and had students map patches of alpine vegetation in the protected lee of large rocks. In the beginning, all of the tiny leafed alpine plants looked the same, but as the mapping proceeded, each plant gained a name and a story. Soon the students saw that a difference in location of just two feet meant the difference between life and death for some plants. And once they understood the wind as the shaper of life in their miniature forests, they started to look around them and see the broad patterns of alpine ecology on the mountainsides all around them. The small world became the big world.

It's unlikely you'll have the opportunity to map tide pools or mountain ridges with your students, but you can use the same principles to design suitable projects for your students. Choose small areas that can fit within the grid frames you're going to use. Try to find places that have large differences in a small area. And match your choice of mapping project to the big ideas in your curriculum. Here are short descriptions of a variety of projects.

Mapping the School Lawn

The many feet that tread the school walkways and playgrounds act similar to the wind in alpine areas. When those feet aren't walking on pavement, they're changing the ecology of the plants and soil underfoot. To illustrate how people shape the environment through their behavior, have the children map chosen areas of the playground.

Divide your children into groups. Have some of the groups map lawn areas that are untrammeled, and have others map areas on the edge of highly trafficked pathways. Provide the children with data sheets that are divided into the same number of squares as the grid frame. Be sure to identify the squares. Also provide the children with a list of the things they should locate on their maps, such as

Stones
Dirt patches
Twigs
Anthills
Pieces of paper
Grass plants
Weed plants

Depending on your interest in plant identification, you might also want to choose three to five of the most common lawn weeds found in your area and teach them to your students before conducting the mapping activity. Most likely you'll have dandelions, plantains, clover, crabgrass, and other easily identifiable plants.

Have the children proceed sequentially from one section of the grid square to the next, adding the information to their data sheets. From this information, create large (preferably 1:1) maps of the different areas. Then examine the maps of the different areas and try to make sense of the differences. Some questions you might ask are

What was the average number of plants in each square? In each area?
In which area were the plants farther apart?
Which area had more dirt patches, stones, and other objects?
Which areas had more weeds and which areas had more grass?
Which plants are able to grow where there's lots of foot traffic?
What are some of the ways in which foot traffic changes an area of lawn?
If we could protect a worn area from people walking on it for a long period, how do you think it would change?

Mowed and unmowed lawns

You can also conduct this activity comparing mowed and unmowed areas of the school yard. The unmowed area will present a much more challenging mapping task to the students because of the height of the unmowed vegetation. It is also harder to represent tall plants on flat paper. However, the comparison provides a vivid illustration of how technology shapes the environment. It will be apparent to your students that there are many more insects, butterflies, flowers, amphibians, and reptiles in the unmowed areas, especially if you choose mowed and unmowed areas that are adjacent to each other.

Grid frames are well suited to mapping the presence of insect life on the forest floor. You should do a bit of prior investigating to determine promising areas before assigning this task to your students. It will also be important for your class to spend time—before or during this project—learning how to identify major classes of forest soil bugs. *Golden Guides* are suitable resources that can supplement more sophisticated insect guides for children. Have the children place the frame on a section of forest floor and gradually peel back, leaf by leaf or bit by bit, the material covering the soil. The style here should be that of an archeological dig, gradually excavating the top levels of duff in order to find the hidden insect life below.

Mapping Anthills

The small-world experience can be accentuated by having children map anthills. Find a sandy area without much vegetation and a number of active anthills. Support the grid on the corners using stones or bricks so it doesn't interfere with the ant activity. First have the children map the features of the area—the anthills, damp and/or dry places, weeds, cracks in the ground, and sticks. Once they have created a map of the area, give them observational tasks on other days. Ask,

> Can you observe routes that the ants consistently follow and show them on your maps?
> Do the shapes of the anthills and the entry locations stay the same or change from day to day?
> What happens if you push one of the anthills to the side? Do the ants rebuild it in the same place?
> What happens after it rains?

There's a general principle to extract from this activity as well. Having a base map of an area that changes will facilitate observation of and long-term commitment to the flora and fauna of an area. Mapping an area is often like staking a claim. The mapping energy expended by the children serves to get them invested in what happens here in the future. The anthills of today can become the conservation areas of tomorrow.

Mapping a Puddle

Puddle tracing

Mapping a puddle can be an easy introduction to the idea of contour mapping, or it can simply be another two-dimensional challenge. If you're doing this as a two-dimensional challenge, it's often a good first step beyond the drawing enlargement activity. Unless there are islands in the puddle, the whole challenge is just to accurately map the shoreline of the puddle—one squiggly continuous line. By adding stones or bucketfuls of sand to the puddle you can manufacture islands and add more dimension and small-world quality to this little water world.

Unless you can find some unusually small puddles, you may have to use more than one grid frame per puddle or have the children move the grid frame. If you use more than one frame, remember to overlap the frames where they abut so you don't wind up with a double-wide edge at the contact point.

Depth profiling

There are two ways to create a depth profile map of a puddle. The simple but time-consuming way is to wait for the puddle to dry up. Map the puddle soon after a rain storm when it is at its fullest. Transfer the outline of the puddle to your paper using the same techniques as described in the previous chapter. Now, to begin the process of depth mapping, trace around the perimeter of the puddle with a piece of chalk or a stick to record the initial shoreline.

Wait a number of hours or until the next day and map the puddle again. If it hasn't rained and water has evaporated, there will be a new shoreline at a lower elevation. Trace around the new shoreline with chalk and map its location. Since the surface of water is always flat, you'll now have a line that is vertically equidistant from your first line. Continue with this sequence of steps as frequently as possible. When the puddle is dried up, you'll have a depth contour map of the puddle drawn on the ground and a contour map on paper. The contour intervals won't necessarily be the same distance apart, but the idea of intervals will be visually apparent.

The second method, creating a depth profile of your puddle without waiting for it to dry up, is a bit more work but is very illustrative. Make sure, however, that you've got puddles with both deep and shallow sections or this activity will be boring. I have always been fascinated by fishing maps that show the bottom profile of lakes and the places to fish for different kinds of fish. This mapping procedure simulates the way these bottom profiles used to be made. In fact, showing children fishing maps will make this process more relevant. Because you need to measure the depth of the puddle at a lot of different points, you may want to add an additional set of strings in both directions to the frame, breaking each square into four smaller squares.

First, map the edge of the puddle as instructed in the previous method. Now, measure the depth of the puddle at each point where strings cross on your grid frame, and record this depth on the corresponding place on a data sheet like the one shown in Figure 5–3. If you're doing this in metric measurement, round off to the nearest centimeter. If you're doing this in English measurement, round off to the nearest half inch.

Now comes the fun part. The challenge is to create the contour intervals by connecting the dots. Choose a contour interval of one inch or two centimeters (see Figure 5–3).

Let's say your greatest depth is four-and-a-half inches. If you're drawing one-inch depth contours, then your greatest depth contour will be four inches. Using a pencil, since mistakes are unavoidable, connect together all of the depths of four inches into an enclosed circular form. If you've got a depth of three-and-a-half inches next to a depth of four-and-a-half inches, you know the contour interval of four inches must go somewhere between them. Similarly, if you've got a depth of four-and-a-half inches right next to a depth of two-and-a-half inches, you know that both the four-inch and three-inch contour intervals have to go between them. After you've completed the four-inch interval, do the three-inch one. This should completely enclose the four-inch interval, and so on. Unless you've got multiple deep

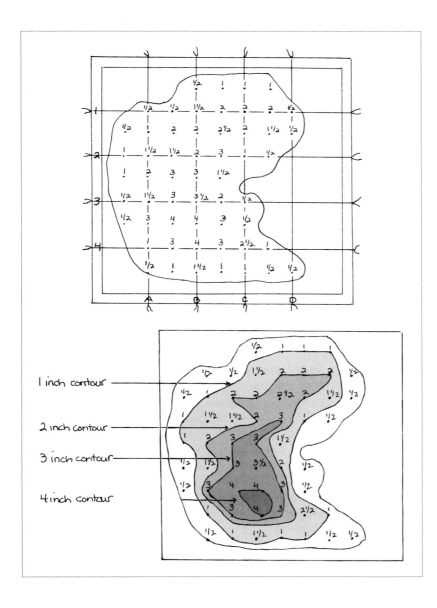

FIGURE 5–3
Outline map of puddle showing depth measurements and contour lines drawn from depth measurements.

spots in your puddle, you should wind up with concentric rings that enclose each other.

Make sure you do a sample lesson with the children on the board to explain this process of *interpolation*. It's initially difficult for students to understand the assumption that if you've got a four-and-a-half-inch interval here and a three-inch interval here, there must be a four-inch interval between them. But once they get the idea, the process of locating the intervals is very satisfying. Once these maps are complete, you can have the children pretend the puddle is a lake and consider questions such as

Where would be the best place to have a beach for young children?
Where would be the best place to fish for deep-water trout?
Where do you think lily pads would grow?
If you had a house on this lake, where would you want to locate it?
Where should you put buoys to warn boats to stay clear?
Where would be a good place to locate a rope swing from a tree growing on the shore? (It wants to be at a place where the water is deep right next to the shore.)

Personally, I enjoy the movement back and forth from academically challenging tasks to imaginative tasks. The play element helps to keep children engaged and also simulates the real-life environmental planning activities they will be drawn into as adults. And the process of going from a large real place to a small map back to an imaginary place builds bridges between children's left hemispheric cognitive structures and their right hemispheric artistic, imaginative capacities. Thoreau said that we must see castles in the air and build foundations underneath them. One is not complete without the other.

THE BASELINE AND OFFSET METHOD: ACCURATE CLASSROOM MAPPING

When I ask children of different ages to draw neighborhood maps, I am often struck by fourth graders' desire to use rulers. It's as if nine-year-olds have arrived at an inner understanding of the importance of measurement and straight lines. Waldorf educators talk about the nine-year-old transition, which heralds a major shift from the more aesthetic focus of the curriculum in the primary grades to a somewhat more cognitive focus in the intermediate grades. From a developmental perspective, therefore, it makes sense to introduce the baseline and offset method of mapmaking around fourth grade. This technique is one of the core tools of making accurate maps, and it can be modified to work in small or large areas.

Though I have already discussed classroom mapping in a previous chapter, I think it makes sense to introduce the baseline and offset method in the classroom to clarify the method and principles. Then it makes sense to transport the method out into the school yard and eventually out into the community.

This technique can be introduced to your students in the context of a common classroom problem. Ask the class to imagine that over the upcoming vacation, the school maintenance staff needs to fix the ceiling or clean and wax the floor in your classroom. All of the tables, shelves, easels, chairs, rugs, and other furnishings will be moved out to the hall during this project. When the work is done, the maintenance staff might not remember how the classroom was arranged. Ask, "How can you make a map of the classroom so the furniture will be put back in exactly the same location?" An emphasis on accuracy raises the stakes from just doing a sketch of the furniture arrangement to using a more sophisticated mapping technique. Once the children are engaged with the idea, here are the steps to take.

Choose a Suitable Baseline

A baseline is a fixed line from which all of your measurements will be made. Once you've completed all of your measurements, you'll transfer the baseline onto a piece of paper and plot locations from it onto your map.

The baseline should be chosen on the basis of ease of use. One natural baseline might be the molding where the wall meets the floor. This will work if the wall is easily accessible all along the side of the classroom so that measurements can be made from the wall outward. However, in many cases heating units and shelves make access to the wall difficult.

Another choice is to choose a straight joint running between two floor boards, or two rows of tiles somewhere near the middle of the classroom. From this kind of baseline, measurements can be taken in both directions. Children can access both sides of a middle line, which makes it easier to have more children collecting data simultaneously.

Divide the Baseline into Appropriate Units

Your baseline essentially serves as a long ruler, so you have to break it up into appropriate units. One handy unit is often the natural dimension of tiles on the floor of your classroom. These tiles are often twelve inches square, which gives you a built-in measuring system. If tiles are not available, then choose an appropriate unit for subdividing your baseline. The unit will depend on the parameters of the mapping project. Since there is a lot to be mapped in a small space in a classroom, a relatively small unit is appropriate, such as one foot or twenty centimeters. It's also useful to choose a unit that can easily be divided in half. Mark off the total length of your baseline with these units.

Draw Freehand Sketches of the Classroom Arrangement or Create Recording Systems for Collecting Data

In preparation for collecting data for the map, the students need to have a system for recording their information. One possibility is to create a grid that identifies each piece of furniture, the offset distance, and the point at which each piece is located along the baseline. While this system is effective, I find that it often leads to numbers getting misplaced and confused. I prefer to have students collect data on a freehand sketch map that shows the approximate location of each object of furniture and attaches the measurements to a visual picture of each thing that is being located. This system can also cause problems if children try to collect too much data in a small space. Nonetheless, I find that having a freehand visual plan helps to keep students oriented and focused.

Demonstrate Measuring Offset Distances

The next task is to measure the distance from the baseline to each of the pieces of furniture in your room. One of the inherent challenges in this process is having children realize that the measurements must always be made at a right angle to the baseline. If you do not demonstrate this process, children will be casual about their measuring.

Take one object, such as the leg of a table, and demonstrate measuring to the baseline using three different pieces of string. Have one string travel from the baseline to the table leg at a right angle, and have the others be off by 10 or 20 degrees. Then compare the three distances from the table leg to the baseline and have children try to explain why they are different.

A simple way to make sure the offset string or measuring tape is perpendicular to the baseline is to use a T square, a book, a stiff piece of paper, or a large right-angle triangle placed at the juncture between the baseline and the offset line to act as a guide (see Figure 5–4).

FIGURE 5–4
Measuring distances using the baseline and offset method. The T-square allows the student to make sure that measurements are taken at right angles to the baseline.

Measure the Offset Distances

Have children work in teams to collect all of their own information, or subdivide the tasks so that each team collects some of the information and then shares it with the other teams. Measuring tapes are excellent for measuring offset distances. If strings are used, they should be divided up into the same units as the baseline. Many small problems may emerge when measuring offset distances. These guidelines may be helpful.

Have children keep the offset measuring strings tight as they collect data.

Remind children to make sure that the angle between the baseline and the offset string is a right angle.

Explain that accurately locating a desk or chair will require more than just measuring the distance to one leg.

Clarify what to do in a situation in which the edge of a table extends out over the legs of a table. (There is no "right" answer; many solutions work equally well.)

Discuss whether it's better to measure the offset from the baseline to the object or from the object to the baseline.

Set the starting point on the offset strings an inch or two from the end of the string so that children have a place to hold on to the string. This will also allow them to easily place the zero point on the edge of the baseline.

Draw a Floor Plan Map of the Room

Here are some guidelines for this task.

1. Large paper (perhaps a 24-by-36-inch piece) will work best for creating these maps. Large graph paper with one-inch squares is optimal; it will be easy to determine a scale of 1 tile = 1 inch. Otherwise, you will need to determine a scale for your map. For guidelines, please see the worksheet titled "Drawing a Map to Scale" on page 150.

2. Locate the walls of the classroom on your graph paper and then locate the baseline in the correct location. Divide your map baseline into the same number of segments as your real baseline.

3. Now locate objects on your map by measuring the appropriate number of units along the baseline and placing a dot there. From this dot, measure the appropriate amount of offset distance and place a dot there. This second dot represents the location of one of the legs of a chair or a table in the classroom. Locate the location of the other legs of this object and then connect these together to create a view from above or symbol of that object.

4. Once you have drawn the location of all of the objects, add a title, legend, scale, and compass rose to your map (see Figure 5–5).

The topography of a classroom doesn't present a terribly interesting mapping challenge and it doesn't get your students out into the natural world. But learning the measuring techniques of the baseline and offset method and learning to handle all of the data are both big tasks. Mastering these skills in a controlled environment will make logistics management much easier once you head for the great outdoors.

USING THE BASELINE AND OFFSET METHOD FOR OUTDOOR MAPPING

I observed a wonderful utilization of the baseline and offset method in a mapping project at the South Brent Primary School in Devon, England. Directly adjacent to the classroom of the fourth and fifth graders was a small,

FIGURE 5–5
Classroom map drawn using baseline and offset method.

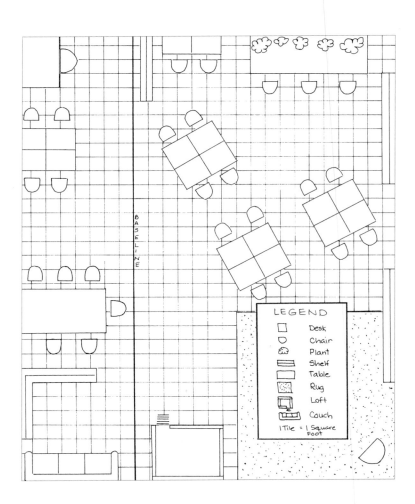

LEGEND

☐	Desk
▽	Chair
⛀	Plant
▭	Shelf
▭	Table
▨	Rug
	Loft
	Couch

1 Tile = 1 Square Foot

oval, cement frog pond about two meters long. The pond had been constructed as a class project some years before. During the spring, summer,
and fall it provided a home for small amphibians, a few water lilies, and
other wetland plants. In the winter, however, it often froze solid, thereby
killing the salamanders and frogs. Each year, it had to be stocked with new
amphibians, or wandering amphibians had to find their way to it.

The teachers and students decided that they could prevent the annual
die-off by creating a miniature bog adjacent to the frog pond. This bog
would provide a place for the amphibians to winter so they could reinhabit
the pond once winter passed. In order to design the bog, they needed a
map of the pond and the adjacent areas so that varying designs could be
drawn and discussed by the class. A group of about six ten-year-olds took
on the project instead of participating in a math unit that they already understood.

The pond was surrounded by a small area of lawn and shrubs that
measured about fifteen meters by eight meters. This was bounded on two
sides by sidewalks and on a third side by a fence separating the playground
from a road (see Figure 5–6). It was only about ten meters from the classroom door to the pond, which made the study area quickly accessible.

The teacher created a baseline by staking out a string that followed the
edge of one of the sidewalks. When the sidewalk stopped, the baseline
continued in a straight line along the same bearing as the edge of the sidewalk. The baseline was marked at one-meter increments and extended
about twenty meters. Students first located the sidewalks, fences, and outside edges of the pond by measuring offset distances, using long tape measures and a T square to assure right angles.

To lay out the shape of the pond, for instance, they found the point on
the baseline that was parallel to the west end of the pond nearest the
wooden fence. This was at 3.0 meters along the baseline. Then they measured the offset distance to the pond edge and found it was .8 meters. At
4.0 meters along the baseline, the near edge of the pond was at .1 meters,
and the far edge of the pond was at 1.8 meters. They kept measuring at .5
meter increments and recorded the following data.

Measurements for Recording Pond Location

Distance along baseline	Offset distance to near edge of pond	Offset distance to far edge of pond
3.0 meters	.8 meters (west end of pond)	
3.5 meters	.2 meters	1.6 meters
4.0 meters	.1 meters	1.8 meters
4.5 meters	.1 meters	1.5 meters
5.0 meters	.4 meters	1.2 meters
5.2 meters	.7 meters (east end of pond)	

Then they determined a scale for the map. Since the area they were dealing with was 15 meters long and the paper they had available was 30 centimeters long, they changed 15 meters to 1500 centimeters and divided to get a scale of 1:50. That meant that 1 centimeter on the paper equaled 50 centimeters of real distance, or 2 centimeters equaled 1 meter. They plotted the location of the sidewalks, fences, and pond edge and then returned outside to locate specific shrubs, rocks, locations of wildflowers, and lampposts. These were then plotted on the maps.

Since this was a special project, these children had to do much of this work on their own. And even though they were mathematically competent, the conceptual movement from data collection to survey map was substantial. I think this is a good indicator that even when dealing with small areas, we have to stay attuned to the conceptual challenge inherent within any task. This task was just within what Vygotsky called the *zone of*

FIGURE 5–6
Survey map of goldfish pond and surroundings created by fifth graders in a British classroom.

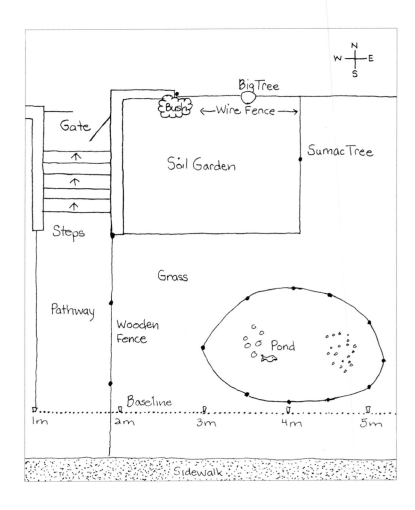

proximal distance for these children. This term refers to the mental realm between what children solidly grasp and what they can be challenged to learn. Being in that zone means that this task stretched these children just the right amount. It was hard but not overwhelming.

When encouraging children to add a new layer of conceptual sophistication, it is important to stay within this zone and not go beyond it. Going beyond the zone leads to being frustrated, tuning out, and feeling a sense of incompetence that undermines learning. Staying within the zone invokes a sense of flow; children are excited to be appropriately challenged.

Further Afield

It is easy to adapt this method for studies of larger areas. As long as there is an available straight line to use as a baseline, the task is relatively simple. When straight lines are not available, it becomes necessary to take bearings using a compass or simple surveying equipment. This is beyond our zone of proximal distance for the moment. However, a long straight road, the edge of a field, a straight line bordering a park, a relatively straight section of stream, or a straight section of trail can all be used as the baseline from which local area studies can emerge. Land use studies, distributions of flora and fauna, and the examination of the ages of buildings can all stem from this use of the baseline and offset method.

These larger studies provide excellent opportunities for collaborative work amongst many students. If the class is conducting a land use study along a section of road near the school, groups of children can be assigned to each block or to fifty-meter sections of the road. In this case, each group collects information on their section and then pools the data to create the final map (see Figure 5–7).

Another method of organization is to have teams of students be responsible for different sets of information. On a street study, different teams could be responsible for:

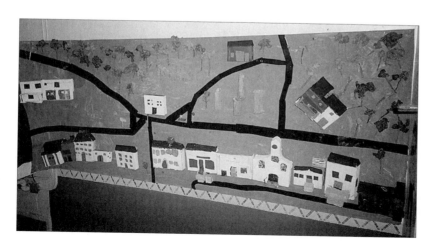

FIGURE 5–7
Three-dimensional wall mural of the village created by third graders in Temple, New Hampshire. The school is located in the lower right-hand corner of the mural.

Sketching the profiles of buildings

Determining the locations of electrical poles

Locating all features that lead to underground facilities (manhole covers, drainage grates, culverts, etc.)

Sketching and locating all traffic signs

Noting sightings and signs of animals

Noting sightings of birds and their nesting areas

Identifying and locating trees

Noting the amount of traffic on the streets

Noting the number of pedestrians and their routes of travel

Watching the information come together and determining the patterns is fascinating. Discuss these questions with the class:

Are the telephone poles all located the same distance apart?

Do different trees grow on different sides of the street?

Which kinds of trees do birds prefer?

Were all of the buildings built around the same time?

How do all the sewer lines fit together underground?

Are all the traffic signs located the same distance from the street?

Understanding these visible patterns provides the foundation for examining broader geographical patterns. For example, why is this city located at the mouth of a river? Why didn't more people settle in this area? What determines the migration routes of birds? These kinds of questions are often posed to children before they've done geographical thinking about areas they have mapped themselves. But once the issues of location, human/environment interactions, and movement have been studied in a small world, students have a jumping-off point for understanding these concepts in the larger world.

CLASSROOM PORTRAITS: SMALL TOWN TOPOGRAPHY

The Parish Maps Project
Rosemary Riddell and David Sobel, Grades 3–4
South Brent, Devon, England

Walk into any news agent's shop in rural England for your copy of the daily newspaper and you'll also find racks and racks of maps. Maps of the village, maps of the nearby cities, the Ordinance Survey maps of the local topography (analogous to our USGS maps), National Park maps, and other maps of various colors and scales. This is in sharp contrast to the relative unavailability of high-quality maps in the United States. It came as no surprise, then, when I discovered that one of the most innovative and exciting projects for mapmaking with children was alive and thriving throughout the United Kingdom in the late 1980s and early 1990s.

"Know Your Place—Make a Map of It!" proclaims the Parish Maps guide (see bibliography), and that's exactly what more than five hundred communities and classes of schoolchildren did. A parish map is a map of the local community made not by cartographers but by the people who live in the community. The "parish," analogous to a township in the United States, is the smallest building block of governmental organization. Exeter, a city of one hundred thousand people in southwest England, has eighteen parishes within its city limits, so here the parish might be analogous to the voting "wards" in American cities.

Locally grown

Parish maps don't just show roads and parks, municipal buildings and schools. Rather, communities are encouraged to make maps that show the features that local people feel an attachment to—favorite trees, pathways, resting spots, sites of local folklore. And maps aren't just lines on paper; they can be artistic expressions in a variety of media. Tom Greeves, director of the project, explained,

> Our primary purpose is to promote and celebrate the importance of the familiar and commonplace to which people are attached, so the whole basis of conservation is broadened. We want to build the confidence of people to express the *feelings* they have about their surroundings and to share these with others. It's as important to preserve the ordinary and familiar things that we care dearly about as it is to preserve the rare and spectacular things that we see only once in a lifetime.

In the classroom

While most Parish Maps projects were spearheaded by community members, numerous teachers took on projects in their classrooms. The encouragement to explore a variety of media has led to the creation of maps of all shapes, sizes, and textures. Tapestries, quilts, photo montages, and three-dimensional models have all been used as map vehicles.

One school in the industrial city of Sheffield based a whole term's work around the idea. Children made maps of their home territory, historical maps, survey maps, spider maps, truancy maps (showing where kids went when they skipped school), and even burglar maps (constructed with the assistance of local police, showing where robberies frequently occurred). These maps were painted, drawn, and created in textile and collage. Parents also contributed to the project, writing impressions of their surroundings. The maps were displayed all around the classroom and gave a wonderful sense of a multidisciplinary approach to learning, crossing boundaries between science, local history, art, creative writing, natural history, mathematics, geography, and craft work (see Figure 5–8).

Special places

Rosemary Riddell and I created a classroom parish map with her class of thirty-five seven- and eight-year-olds in the autumn of 1987. Our objective was to create a map of children's special places in the parish of South Brent, a community of a few thousand people nestled in the hills on the edge of Dartmoor National Park.

We started by having children draw a map of the special places in their neighborhoods—favorite places where they played with friends or by themselves, secret places, and places to explore. So children's maps included their private dens, tree forts, private pathways, and the ghost house (an abandoned house suspected of harboring spirits) on the moor. Next, I put children in neighborhood groups and had them work on maps of their local areas, trying to make composite maps of everything they knew about

FIGURE 5–8 *Clyst Hydon map created by students working on a Parish Maps project in their British classroom.*

their neighborhoods. One group of children made a composite illustration of Mucky Lane, a slightly eerie farm lane near their homes. On the map one child wrote, "When I walk down Mucky Lane in the fog it feels like someone is creeping up behind me." Another group identified an important bridge where they liked to play, and one child wrote, "This is the hole in the bridge that my dog sticks her nose up through."

On the wall of the classroom, I mounted a four-by-five-foot map of the parish, including the river Avon, the major roads, and the railroad line. We located where each child lived, and then cut up their maps and started to add pieces of their maps to the class map. During this process, we identified some of the special places around the center of the village. I did this by passing out a worksheet that asked children to list beautiful places, scary places, friendly places, junky places, exciting places, quiet places, and dangerous places. Our objective was to get beyond the well-known and conventional sites and tap into the real emotions that connected children and place. We wanted the good, the bad, and the ugly.

As a result, we discovered all kinds of idiosyncratic places that wouldn't make it into even the most detailed guidebooks. The churchyard at night, the creepy village toilets, and underneath the railroad bridge were all scary places for the children. Lots of children talked about Fat Man's Trouble—an extremely narrow public footpath between two stone walls that circumferentially challenged people wouldn't fit through. One child identified the bridge where he played Poohsticks, and other kids talked about the valley where the Beast of Dartmoor, a local folklore figure, allegedly lived. We had individual children or pairs of students work on small pictures of these places. Gradually, over the weeks, we built up a collage of the parish that captured the children's mythology and personal lore. As opposed to the hard-edged feeling of conventional maps, ours developed a lived-in feeling.

Teaching about scale

To help the children understand the idea of scale, I created the Incredible Shrinking Machine. I took a picture of a child's house that was much too large for the scale of the class map and asked the children if we should attach it to the map. "No!" they all chorused, "it's way too big." I took the picture, slipped it inside the envelope, had the children say a shrinking incantation with me, and then opened the envelope. Inside was the same drawing, but it was half the size as before. We tried it again, and they all agreed that it still didn't look right. We shrank it again and again until we got it down to the right size for our map. This magic was accomplished through predrawn sketches of various sizes that were all in the envelope from the start. Nowadays, of course, photocopier reductions would make this easy to do. But the fact that the children could tell that these were all real pencil drawings made the reductions seem even more magical.

After this, whenever children brought a contribution for the map that was too large, I said that the drawing needed to be put in the shrinking machine in their minds. They would return to their desks and work on the same drawing, only smaller. The "Incredible Shrinking Machine" activity served as an effective transitional metaphor to convey the idea of scale to them qualitatively. We never established a scale or even used the term, but the children developed an intuitive sense of what was about the right size and what was inappropriate. This helped to give the children a kinesthetic understanding of scale upon which a mathematical understanding could later develop.

We filled the map out by cutting out pieces of cotton, fleece, and corduroy to represent all of the grazing fields around the parish. The map developed a patchwork look that was similar to how the landscape appeared from on high. We made small vehicles for the roads and framed the upper edges of the map with children's paintings of moor landscapes. It was geographically accurate *and* a work of art.

The true success of the process, however, was in the validation of the children's unique perceptions of the places around them. When asked to describe a favorite public footpath that was shaded and bordered by a tiny stream, one boy pensively offered, "It feels like there's a toad in every corner." Another boy described an abandoned house where he played, saying with barely contained fear, "I saw a skeleton head peeping up over the wall." The map was like an X-ray of the parish, showing the hidden skeleton of play experiences, adventures, and sadnesses that give a place its character.

A Relief-Model Project
Mary Hayward and Sherry Bartlett, Grades 3–4
Westminster West, Vermont

The Westminster West relief model was presented to the community at a ceremonial June celebration, complete with a bonfire. On this long, warm summer evening, it was hard to remember the beginning of the project during the mud and ice storms of March. The project was conceived by parent and geographer Sherry Bartlett and was created by the third- and fourth-grade students of teacher Mary Hayward in the tiny Westminster West school. Both had been inspired by the Parish Maps model from England.

The students started by researching as many subjects about the town as they could think of. Some of their findings went into the map while some information found its way into a companion project, the Westminster West Atlas. Topics included some uses for old barns; people's occupations and animals; and the locations of beaver ponds, springtime flowers, cemeteries, and houses. The final model was constructed on a four-by-eight-foot sheet of plywood, using chicken wire for the contours, and it was covered with papier-mâché and grocery bags. Tiny sheep and cows populated the fields; miniature apples hung in the orchards. Fields were illustrated in light green; forests were a darker green. Paved roads, dirt roads, and trails were all shown in different media. Houses, farms, and barns were exactly located, and annotations told of the history and life of the village, mostly from the children's perspective.

Some of the stories were true—"This is whair my dogs are buried." Others tended toward the fantastic—"Impenetrable quicksand swamp. Beware of strange life forms." Students even included a gopher hole that supposedly stretched all the way from Westminster to Westminster West. But whatever was included had to be voted on in strict accordance with democratic procedure, and heated exchanges did sometimes result.

One subject that generated strong feelings was the discovery that the Pinnacle, long a favorite hike for locals and the highest point in Westminster, was no longer accessible. Apparently, the owner of the land across which hikers needed to pass to reach the Pinnacle had declared the property off limits. Consequently, the Pinnacle on the students' model was swathed in red paint, with barbed wire surrounding a flag that bore the inscription "Is it fair to say no to people walking up the Pinnacle?" The legend "mountain snatchers" on the model house at the base of the hill left one in no doubt as to the children's answer.

6 | UP, UP, AND AWAY
Ages Eleven Through Twelve

BEYOND THE HORIZON

In elementary school, I was always one of those good students—the kind teachers like and other kids make fun of except when they want to copy homework. I knew all the state capitals, and I liked coloring in those maps of the different regions of the United States. But I always had trouble with indigo. Indigo was one of the major products of South Carolina. First of all, I was a little confused by this product idea. Each state had two or three of them, which didn't seem like quite enough. Is that all they made in that state? And in our suburban commuter town, there didn't seem to be *any* products being manufactured. Even by fifth grade, the whole idea of livelihood and production was still a little beyond me. My mother sold houses and my father was a travel agent, but houses and airplane tickets weren't really products, were they?

Cows and dairy products I understood. I had a glimmer of an understanding of changing land use patterns because the big dairy farm where we used to go to touch the electrical fences and snoop around the milking barns was sold. It became the corporate headquarters for a major chemical company. But indigo still puzzled me. How could a color be a product? Or how could a plant that made a color get bought and sold? Still, when I was asked on the test about South Carolina's major products, I dutifully wrote down *indigo, cotton, tobacco,* and *manufactured goods.* And though I never really understood what this meant, there was a piece of me that wanted to understand, to get a sense of how money and people and places all fit together.

Indigo symbolizes for me the abstractness and developmental inappropriateness of much of elementary school social studies and geography. The intent of curriculum designers and teachers seems right, but the methods don't work. Expanding students' horizons and understanding one's place in a big world are important objectives. But unless the learning is tangible and grounded, the information provided to students will be like water off a duck's back.

A few years ago, I reconstructed an autobiography of my interest and involvement with maps since childhood. I tried to identify one or two memories relating to maps from each year of my life. I wanted to be able to see both the developmental variations at different stages in my life and the recurring regularities. The changes from ages ten through fourteen are revealing. At ten and eleven years old, these memories stand out.

Age 10

My two favorite board games were Square Mile and the Lone Star State. In Square Mile, you developed tracts of land three-dimensionally, developing the land with little roads, bridges, and buildings that came with the game. It was a notch or two beyond Monopoly and a precursor to Sim-City. In the Lone Star State you moved your marble through a molded plastic landscape that had mountains peaks, slides, short cuts. I liked any game that took place in a simulated landscape.

Age 12

I was astounded with the cool nonchalance and savvy of Ray the Good Humor man. When I asked for a Toasted Almond ice cream bar, he opened the door in the freezer compartment of his truck, reached into the voluminous darkness without looking, twisted his arm two or three times, and voilà, pulled out my order. It all happened so effortlessly. I realized that he had a three-dimensional map in his head and hands of the interior of the freezer, and I felt a desire to have the same kind of knowledge.

Both of these memories suggest my persistent fascination with small worlds. I liked operating within the simulated world of the game board and wanted to be able to have a map inside my head that would help me to understand the small world inside the Good Humor truck. But by age twelve, a new kind of interest was creeping into my consciousness. I wasn't losing my interest in small worlds, but I was adding on a new fascination, the way a nautilus grows by adding a new chamber to its shell.

Age 13

I started a collection of state highway maps. I clipped the ads from *Life* and *National Geographic* and sent away to the Tennessee Tourism Office and the Alaska Vacation Planning Center for my packets. Once they arrived, I pored over the beautiful, unusually large maps and started to plan adventures.

Age 14

My mother hit the nail on the head when she bought me a gigantic world map. Wallpapered to my bedroom wall, it was about eight feet by twelve feet and was the source of hours of fascinating games. "Find Novosibirsk," I'd say to my brother, and he would have thirty seconds to look through the wide expanses of Russia to locate it. I can still feel the expanse of Siberia stretching vastly across the wall above my desk.

My perspective was expanding from the small world to the big world. Maps were starting to feel like books to me. I could spend time with them and enter into the landscapes they represented. They ignited my imagination and served as a different kind of magnet, pulling me toward far-off places.

It's valuable to recall the neighborhood maps of children this age. By twelve, about half of Western children are drawing mostly aerial view maps of wider and wider terrains. By fourteen, almost all children opt for this perspective. This illustrates the importance of moving up, up, and away—but not too quickly. By the upper elementary grades, the standard curricular emphasis is beyond the horizon. Fourth graders in Vermont and New Hampshire are supposed to study state history and geography. The state capitals, national and world maps, and the geography of the thirteen colonies are standard issue. But many of these youngsters are still somewhat tied to a small-world consciousness. I think New Hampshire geography should be a fifth- or sixth-grade topic, but that's not going to change anytime soon. So what are we to do?

The solution is to take on the big world in a small-world fashion. We can teach about the long ago and the far away in a hands-on, minds-on fashion. Take the topic of state products, for instance. I know a fifth-grade teacher who has children study regional maps by creating symbolic collages. Each state on the map has to be represented by a collage of two or three things made from a material that comes from that state. Montana, for instance, would be covered with pennies to represent copper mining and tiny crackers to represent wheat. Georgia might be represented with fruit stickers from peaches and unshelled peanuts.

A fourth-grade teacher responsible for teaching U.S. geography puts up a large map of the United States on a bulletin board and has children bring in things that come from each state. The objects are arrayed around the edge of the bulletin board with a string from the object to the state. There are postcards that originated in North Dakota and Arizona, Saturn owner's manuals from Tennessee, pint jars of maple syrup from Vermont, and five-pound bags that contained potatoes from Idaho.

At the same time, there are increasingly more sophisticated ways to map the local environment. Students can start to grasp more complex surveying techniques, are thrilled by the idea of orienteering, and have the conceptual ability to start dealing with the puzzle of representing topography through contour mapping. A balance of the near at hand and the far-flung will make for a healthy curriculum in the intermediate grades.

"Oh, cool!" was my seven-year-old son's reaction on his first encounter with a raised relief map. The map hung on the wall at a car rental agency in San José. Though his map-drawing skills were still fairly primitive, he immediately grasped that this was a map that actually showed the up-and-down shape of the land. Costa Rica's spine of volcanoes, a sliver of which we could see from the house we stayed in, was finally clear to him. It brought back the feeling of my early encounters with relief maps. I loved running my fingers over the shapes of the mountains and down the valleys of rivers, trying to absorb the landscape through my skin.

Raised relief maps begin in the sandbox as early as children make roads and hills for their cars and trucks. In the classroom, sand trays and stream tables provide experimental geography opportunities. What shape of river do we get when the slope is steep? When it is gentle? What do you need to do in order to make a waterfall? What happens when a river meets the ocean?

The next step is to use the same principles of active education to create raised relief maps of larger areas. Though I will describe the process of mapping a country in this chapter, keep in mind that this process works at many different scales. You can create raised relief maps of imaginary islands from storybooks, of towns, or of river valleys. A fourth-grade teacher in Townshend, Vermont, centers her required study of state history and geography around the relief maps of Vermont created by groups of four students. The experience of shaping the topography provides a sense of familiarity when the scope of the map stretches beyond the known world of the child.

One of the best raised relief map projects I have ever observed occurred in a sixth-grade classroom in a Waldorf school in Dartington, Devon, England. The Waldorf curriculum is attentive to developmental stages, and their use of mapmaking is somewhat consistent with my convictions. Though Waldorf teachers don't do much with maps in the first three grades, they initiate a study of geography in fourth grade by having children draw maps of the school's neighborhood and the route children follow from home to school. This opens out into a study of the community and the geography of the state later that same year. The class I observed in Devon was involved in a theme study of the history and geography of the United Kingdom, and the mapmaking component took place twice a week for about six weeks. You can use the steps this teacher followed to create a raised relief map of a country in your own classroom.

Choosing a Suitable Base Map and Imposing a Grid on It

This classroom teacher chose a base map of the country from the classroom atlas that all children owned. The original map was about eight inches by ten inches, and the teacher imposed a four-by-five square grid so that each section was approximately two inches by two inches. Each of the fifteen children in the class was assigned one of these sections. Since three of the sections wound up to be only ocean, there were enough to go around and some students had to do more than one section. The teacher wrote the name of the student responsible for each section on the original map.

Gridding Up Each Section of the Base Map

Each child enlarged his or her section of the country from the original two-by-two-inch square up to a ten-by-ten-inch square. Each child imposed a smaller one-by-one-inch grid on the small map section and subdivided the ten-by-ten-inch square into a five-by-five-inch grid (see Figure 6–1). Thus

FIGURE 6–1
Four steps in the creation of a 4 foot by 8 foot raised relief map of the British Isles. (A) Grid laid over map of British Isles; (B) Individual section gridded up by student; (C) Building up the landscape of individual sections with plasticine; (D) Assembling all the sections on a table in the center of the classroom.

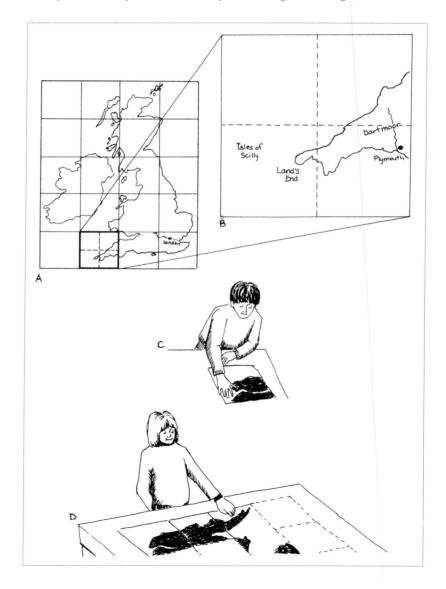

the four squares on the small section corresponded to four squares on the enlarged section. Using the gridding process, each child then transferred the boundaries of the coast and the locations of rivers, mountains, and cities onto his or her own enlarged section.

As students completed their enlarged sections, a wonderful piece of impromptu cooperative learning emerged. The teacher required each child to check the points at which coasts and rivers left one section of the map and traveled onto the neighboring section. In many cases, students found that their coastlines didn't match and their rivers didn't flow into each other, so they had to do some negotiating. (Some students tried to argue that there had been a recent major earthquake.) Students working on neighboring sections had to check their enlargements to see how to adjust both maps so coastlines and rivers would meet. Instead of the teacher being the judge of what was right or wrong, the involved parties had to reach their own agreement. Before a student could go on to the next step, all neighboring parties had to be in agreement.

Researching and Learning About the Landforms of Each Area

The children were each responsible for researching the geography of the section they were working on. Using classroom and library resources, each child had to find pictures of the geography of the section and identify ten significant geographical features of the area. In the section that included most of Devon and part of Cornwall, these features included such places as the high rolling moors of Dartmoor and Exmoor, the river Dart, the beaches of Torquay, the south coastal cliffs, the harbor at Bristol, and the north Devon coastal marshes. The teacher wanted the children to have a palette of geographical images in their minds to provide the foundation for the next step.

Building Up the Landscape of Each Section

Now came the fun part. Each child was given a large supply of plasticine to build up the topography of the section. The drawn map was used as a base map so the plasticine was laid directly on top of the map. Everything was covered with a thin layer of plasticine, and then higher areas were built up. As significant features on the base map were covered up, students used toothpicks to identify locations, or they sketched locations into the soft plasticine. Students realized that to create rivers they had to build up the areas around the rivers in order to have enough depth to carve the rivers down into the landscape. By shaping the landscape with their hands, they were gaining a kinesthetic experience of the geological forces of land formation and erosion.

Developing a Vertical Scale

On raised relief maps, the vertical scale is different from the horizontal scale. Even on a large map of England, such as the one in this classroom, the horizontal scale was approximately 1 inch = 5 miles. If this were used for the vertical scale, then Snowdon, England's highest peak at 3560 feet, wouldn't even be one-quarter inch high. Instead, an appropriate scale for this kind of raised relief map might be 1 inch = 2000 feet so that Snowdon would end up being about two inches high. The scale could even be more exaggerated so as to accentuate the height of the hills and mountains and the depths of the river valleys, though this requires a lot of plasticine. One strategy is to determine a scale ahead of time and have children build up their mountains in accordance with the scale. One effective way to do this is to mark toothpicks with the correct height of the mountains and then build the landscape up to that level.

In the Dartington classroom, the teacher let a qualitative scale emerge as the children worked independently on their sections. After examining many different sections, he called the class together and said, "Nicholas has Dartmoor in his section where the highest hills are two thousand feet high, and he's got those hills two inches high. But Amanda has the peaks of the Lake District, which are about three thousand feet high, and her peaks are one-and-one-half inches on the map. What's the problem here?" A consideration of whether to make the scale larger or smaller involved an assessment of the available material. If 1 inch = 1000 feet, then everybody with mountains was going to have to add a lot of plasticine to their sections. On the other hand, without much height to the mountains, there wasn't enough drama.

This kind of push-and-pull discussion is valuable for students this age because it opens up the notion that all maps are manipulated in specific ways to make certain points. The upside-down maps of the world that put the southern hemisphere on the top and the northern hemisphere on the bottom are good examples of how maps can manipulate our thinking. Denis Wood, in *The Power of Maps* (1992), contends that all maps are designed with a purpose in mind and that if we are blind to the purpose then we can be easily persuaded to take the cartographer's perspective. These kinds of discussions about choices of projection and representational style become appropriate in the upper grades and emerge naturally when children share in making the decisions about how the map will work.

Adding Rivers and Cities

Once the topography was in place, the rivers were carved into the landscape and cities located. In this classroom, the rivers were represented with blue plasticine and the cities with red. Before rivers were laid down, the teacher evaluated their slopes by squeezing eyedroppers full of liquid into their upper reaches to see if water flowed all the way down to the ocean. If not, students had to adjust the pitch of their river courses. And rather than placing cities on top of the green plasticine of the landscape, the students excavated a bit of the surface first to represent the processes of digging foundations and dredging and filling that happen during city construction.

Putting It All Together

Throughout the process, the children had all been working patiently on their own sections. They had to check with their neighbors and adjust their scale, but no one really had a sense of the whole picture. As the sections were nearing completion, the teacher brought in a four-by-eight-foot piece of plywood and supported it on desks in the center of the room. Then, starting with the northern reaches of Scotland and the Orkney Islands, one child at a time brought up a section and fitted it into place. There was a quiet, held breath kind of feeling in the classroom as the disparate pieces flowed together gracefully. And when the tip of Cornwall was grafted on, a "wow" of satisfaction filled the room. The children's pride was tangible in the air. The size of the map and the handcrafting of the landscape gave the students a grasp of the structure of their homeland. Map projects done in sections and then assembled together provide a metaphor of cohesiveness that contributes to weaving the social fabric of the classroom group.

This approach can also work well two-dimensionally without the use of plasticine. I have seen even larger maps of the United States done in a similar fashion on large sheets of paper. The Vermont teacher mentioned previously has small groups work together on smaller relief maps of Vermont (see Figure 6–2). If you want to be even more industrious, however, try the map project described in the next section.

FIGURE 6–2
Raised relief state maps created by fourth graders in Townshend, Vermont.

A large map of the United States painted onto the asphalt surface of the playground serves as the ultimate tool map. Making the map is valuable, and the finished product becomes a permanent learning tool for games and various aspects of the social studies curriculum. And don't limit your horizons. Consider a map of North, Central, and South America to foster an understanding of geographical and cultural diversity. In England, a map of Europe on the playground would be a step toward transcending nationalism and embracing the European community.

Large maps can be made of smaller areas as well. For example, a playground map of the state of Colorado would enhance the teaching of Colorado state history and geography in the upper elementary grades. A map of the major bioregions of the South might be the choice for a middle school with an environmental focus in Georgia. The fourth-grade teachers at the elementary school in Newbury, Vermont, have painted a twenty-foot-long wall map of the Connecticut River and its tributaries as part of an ongoing study of their own river ecosystem. Having the map on the wall gives children a sense of their place in the watershed. Here, I will focus on the steps of making a map of the continental United States, but the procedures are basically the same for the creation of any large map. If you want to include Alaska and Hawaii, you have two choices. To put them in their correct geographical positions, you'll need an extremely large expanse of asphalt. Otherwise, you'll have to stick them in those self-contained boxes so popular in atlases.

Choosing an Appropriate Site

For a playground map of the United States to really work, it has to be big enough for a group of ten or more children to play games on. Twenty feet won't do; forty feet from coast to coast is a minimum, and sixty or more feet is even better. Make sure you've got a large enough area to do this.

Orientation is the next challenge. For the map to be geographically useful, it needs to be oriented correctly to true north. Don't make the mistake of fitting the map to the slab of playground asphalt without paying attention to direction. It doesn't make sense to have the sun rising over the Gulf of Mexico and setting into Canada. If the playground slab does not have a true east/west or north/south orientation, the map will appear askew in the beginning. But the advantage of being able to stand in California and look east toward the rising sun coming up over the Rocky Mountains will outweigh the problem of not having the map fit perfectly into the center of the playground slab.

Creating a Map Puzzle

One way to transfer the map outline onto the playground surface is to use the gridding up technique described earlier in this book. You can also try a technique that was used at one of the local elementary schools in Keene, New Hampshire. It has the advantage of being simple enough so that some parts of the map can be done by primary-age children. (Keep in mind, however, that I consider this project best for fifth graders and above.)

To use this second technique, choose a base map of the United States to work from or an atlas that has all of the states represented in the same scale. Do some initial calculations to figure out how big the individual states should be so that the assembled pieces of the puzzle and the final map actually fit on the playground. You can use a state such as Kansas as a measuring device. Knowing that the east-to-west distance across Kansas is about one-tenth of the distance across the United States from the eastern tip of Maine to the west coast of Washington, you can scale your map accordingly. If you make Kansas five feet long, then the whole map will be about fifty feet from coast to coast.

To make the templates to trace around on the playground, you can take two approaches. First, if you want to involve all of the students in your class, you can have the children carefully trace the outline of each state onto paper. Then, using an overhead projector, project each image onto large sheets of stiff paper on the wall and trace around the outlines. Or, if you have an opaque projector, you can simply project the map image on the wall and trace around the individual states. The drawback is that this approach makes it harder to involve all of the children simultaneously.

In either case, it is crucial to remember to maintain the overhead or opaque projector at the same magnification and distance from the wall for each state. Otherwise, the scale for states will be different and the puzzle pieces won't fit together. Make sure that no individual state is too big in its original form to fit on the projection surface.

Once all of the states have been traced, you've got a set of puzzle pieces. It's a good idea to lay out the map on the gym floor to make sure that all of the pieces fit together correctly. It's a lot easier to make corrections at this point than it is once you start to put paint to playground.

Transferring the Puzzle to the Playground

To assure correct orientation of the map, it is necessary to lay down a set of north-south and east-west lines. Using a compass to determine north, snap a north-south chalk line in the approximate center of the area to be painted for the map. At a right angle to this, snap an east-west chalk line. Since the southern border of Kansas is oriented east-west and Kansas is located near the geographic center of the United States, you can begin the transcription of the borders of the state by aligning Kansas on the east-west line with its southeast corner at the point where your north-south and east-west lines meet. Trace around Kansas with chalk, remove the puzzle piece, and then lay out all the states that border Kansas in their appropriate positions. Trace around these with chalk and then remove them.

The next step is to make these chalk lines and the cardinal direction lines semipermanent by painting them with faint dotted lines. You should only chalk out as many states in one day as you can also paint that day. Otherwise, the chalk lines will be eradicated by foot traffic and weather, and you will have to redo them. On subsequent days, lay out the rest of the states and lightly paint all of the borders until all of the states are completed.

Painting the Map

Once the outline is completed, the hard work is done. Now comes the fun of painting in each state. Since you may want to add in rivers, mountains, lakes, state capitals, and the like on top of the painted base map, it's best to use light colors to differentiate the states. Since you don't want contiguous states to be the same color, the number of colors you'll need will be determined by the state that has the greatest number of bordering states. (This is a great geography math problem. Explain the paint problem to your students and let them figure out which states have the greatest number of bordering states.) Once you've determined the number of different colors you need, let a couple of students figure out the color pattern. Suggest that they start with those states with lots of bordering states and work outward from them.

Then the painting begins. Don't try to do it all in one day, and make sure to have teams working in different regions at the same time. The paint will take time to dry, and it's easy to make a mess by leaning into not-quite-dry Kentucky while painting Tennessee. Recruiting parent volunteers so that you can have painting teams of one adult and two or three students will maximize tidiness and help to keep paint off clothes and the adjacent four-square courts.

Embellishment is up to you. Since political boundaries don't always reflect ecological boundaries, I suggest adding mountain ranges, rivers, and major lakes. What map of the United States would be complete without the Mississippi River and the Great Salt Lake? By choosing lighter colors for the states, you can use darker colors for these features. It's attractive to include bordering oceans and to indicate the location of our North American free-trade partners. You might want to leave off the names of the states to encourage children to learn them, but including the location and names of the state capitals will win the hearts of fact-oriented parents and school board members. Don't forget a compass rose.

Is This Kansas, Toto? Using the Map

This book isn't about map reading or map use, but I can't resist a few suggestions. State recognition games on the big map serve as a great motivator to learn all of the states and the capitals—an appropriate objective by the end of elementary school. Using a basic tag format, you can play "Safety Is." The person who is It calls out, "Safety is Wisconsin," and anyone who isn't in Wisconsin can get tagged. There are numerous variations: "Safety is any state that begins with *M* . . . any New England state... any state west of the Mississippi . . . the state whose capital is Tallahassee." If you don't know these locations, you can't play the game, so the map provides the raison d'être.

The relative size of states can be explored by seeing how many people can stand on Rhode Island versus how many people can stand on Oregon. Answers to questions such as, Is it farther from Chicago to Disneyland or Disneyworld? can be determined by comparing the number of child steps between these destinations. Classroom simulations that involve traveling across the United States on maps can be conducted on the playground. Thoughtful teachers will devise new uses each year.

The playground map serves as a metaphoric bridge for the students. Being able to compare the physical experience of the few baby steps from New York to Boston versus all those giant steps it takes to get from New York to San Francisco allows students to grasp the magnitude of these United States. As younger students learned to emotionally inhabit their homes, neighborhoods, and communities, older students grow into a relationship with bioregions, the nation, and eventually, hopefully, the world.

GOING BEYOND RIGHT ANGLES

A Note on Compass Use

I have chosen to emphasize mapping techniques that do not require compass use with elementary-age children for a number of reasons. First, I find that it is difficult for children under ten both to use a compass accurately and to understand how a compass works. To use a compass well, children must be able to take a bearing that is accurate within about five degrees. Many fifth and sixth graders will not have the patience and maturity to do this. Second, most teachers don't have a good set of classroom compasses available for their use. Most inexpensive school compasses are not suspended in liquid, don't have sighting aids, and aren't large enough to be useful for mapping. Using them for these purposes just leads to frustration.

Some of the techniques described in the following section will be enhanced with the use of compasses. If you want to invest in good compasses and teach your students compass use, there are a variety of good teaching materials available. I recommend the *Map and Compass—Orienteering* book, published by Orienteering Services of Canada, for upper elementary teachers.

Plane Tabling

The next step of mapping sophistication beyond the baseline and offset method is a technique called plane tabling. One form of plane tabling allows children to identify the locations of objects at a distance without the use of a compass or tape measure. Other surveying techniques require the use of both of these tools. Additionally, you may want to use an enlarged circular protractor (Figure 6–3) for some of the work described below.

A map of a large area, such as the school grounds, a cemetery, or a nearby town park can be made through plane tabling. This technique requires that the students take two sets of sightings on objects around the periphery of the area, plot the sightings on paper, and then determine the locations of the objects by seeing where the sighting lines intersect. (Yes, this is a bit mathematically complicated—see Figure 6–4 to get a visual picture of the process.) After you have a personal sense of the following steps, demonstrate this process indoors before trying it outdoors.

1. Pin a large sheet of plain drawing paper (approximately eighteen inches by twenty-four inches) to a large drawing board.
2. Mark a straight line on the ground near the center of the area to be mapped. The length of the line depends on the size of the area. Ten feet is usually suitable for an average-sized playground. Larger areas will require longer baselines.
3. Either estimate the length and width of the area to be mapped or take quick measurements so that you can determine a scale for the map.

FIGURE 6–3
Protractor for use in simple transit and measuring tape mapping technique.

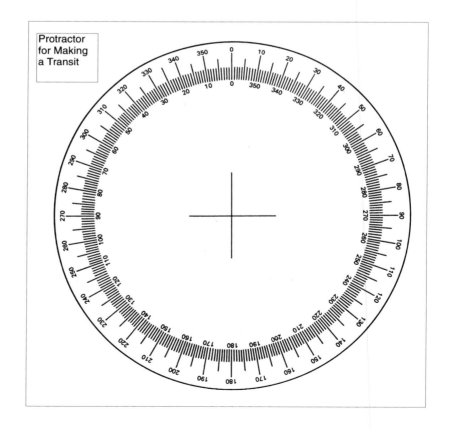

4. Draw a straight line in the center of the paper that corresponds to the straight line on the ground. Determine the length of this line by translating the ten feet of the line on the ground down to the appropriate scale on paper.

5. Place the drawing board on a small table or other support so that one end of the line on the paper is directly over one end of the line on the ground. The bottom edge of the drawing board should be parallel with the baseline. Other things that will serve as tables include bushel boxes, cardboard boxes, stepladders, or overturned garbage cans (see Figure 6–4).

6. Make a simple sighting instrument by attaching a small nail or tack to the center of each end of a short wooden ruler. The nails

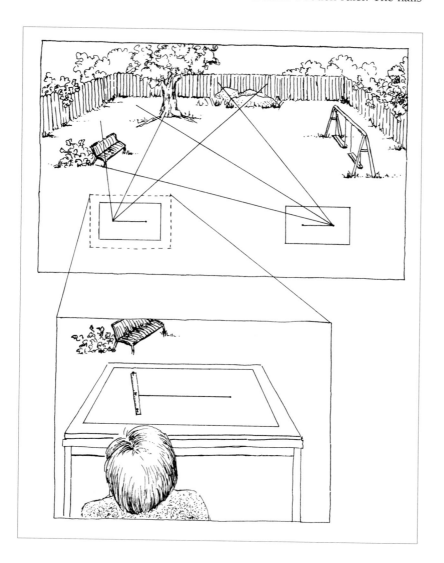

FIGURE 6–4
*Using the plane-tabling
method to map the
school grounds.*

allow the students to accurately aim at objects at a distance. Drill a small hole in the center of the ruler and attach it to the end of the line on the paper with a pushpin.

7. Once the sighting instrument is attached to the drawing board, aim it so that it is pointing to one of the boundary features, such as the corner of a stone wall, a large tree, a stop sign, or a corner of a building. While one child is sighting, have another child place a pencil mark on the sheet at the center of the far end of the sighting instrument. This mark should be numbered. On a separate sheet of paper, notate the number and the object being sighted.

8. Now turn the sighting instrument and aim it at a variety of points around the periphery of the whole area. Try to choose objects that designate corners of the area to be mapped. At each point, make sure that the direction in which the sighting instrument is pointing is marked on the map, that the mark is numbered, and that the object aimed at is denoted on a separate piece of paper. A total of six to ten features will usually provide enough information.

9. When you have a complete set of sightings, remove the sighting instrument and join the points on the paper to the correct end of the baseline with light pencil lines.

10. Move the drawing board and table to the other end of the line marked on the ground. Fix the sighting instrument to the other end of the line on the paper.

11. Complete another set of sightings on *exactly* the same points as before, following the same procedure. When this second set of sightings is completed, connect the points to the other end of the baseline.

12. Back in the classroom, extend out all of the lines and find the intersecting points. For instance, extend out the sighting lines from both ends of the baseline to the first object, the big tree. Where these two lines cross indicates where the tree is located. Find the intersecting points for all pairs of lines.

13. Once you have all of the intersection points, connect them together and remove the lightly drawn sighting lines. The lines connecting the intersection points constitute the boundaries of the area you are mapping. You may want to symbolically or pictorially represent the boundary points in some fashion.

Alternative Approaches

The process described above is sometimes called *triangulation*. If you'd like to introduce a bit more mathematical accuracy, you can accomplish the same task as the previous method with the use of a circular protractor or a compass.

1. Mount the circular protractor (see Figure 6–3) to a square piece of plywood.
2. Orient the protractor so that the 0–180 axis points toward two of the corners of the board, and draw a straight line from one corner of the board to the other through the center of the protractor.
3. Create a sighting instrument similar to the one described in the previous method. Attach the sighting instrument to the center of the protractor on the board (see Figure 6–5).
4. Instead of using a baseline, take your sightings and measurements from one center point. Choose an identifiable location for this point, such as a sidewalk corner or a storm drain. Place your table directly over this point and choose your first boundary object to aim at.
5. Choose the six to ten objects on the boundary of the area you are going to map. Create a quick sketch map of the area showing the location of your center point and these boundary objects. Don't worry about scale because this is simply a rough draft to use for recording information.
6. Orient your circular protractor so that the 0–180 line points directly at your first object. Now, rotate the sighting instrument and take sightings on the other boundary objects. Read the direction bearing from you to the object off the protractor and record it on your sketch map.
7. When you complete a circle of sightings, measure distances from your center point to the objects. Record these distances on your sketch map. Now you have a complete set of information.
8. Draw a map by transferring the information from your sketch map onto a scale map. Use a small plastic circular protractor to transfer the information to paper.
9. Determine the scale for your map and then choose a center point on the paper to correspond to the center point you used outside. Choose an approximately correct location for your first object on the paper and then orient your protractor so that the 0–180 line points from the center to the object. (Since you are not using a compass that is oriented to north at this point, you don't have to worry about true directions.)
10. Translate the information by reading the bearing off the sketch map and finding the same bearing on your protractor. Translate the actual distance outside to the scale of your map. Once you've located all of the boundary points, connect them together and erase the bearing lines. Now you have the boundary of your area.

Using a compass and a measuring tape

The procedure is essentially the same except that with a compass you are taking true bearings rather than taking bearings in reference to a fixed

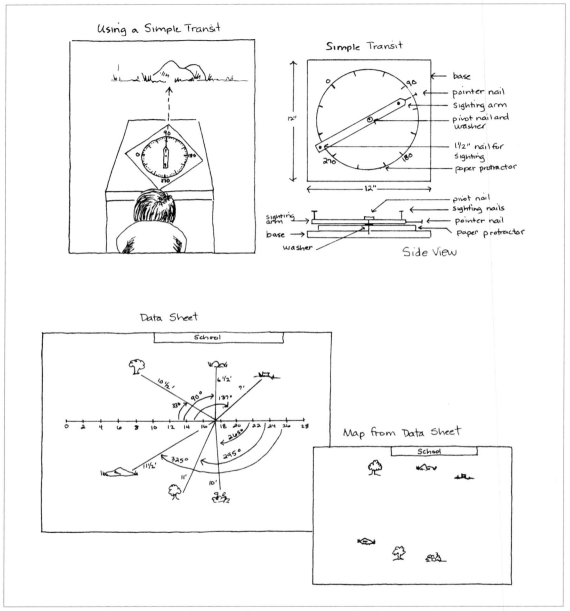

FIGURE 6–5 *Mapmaking using a simple transit and measuring tape. (A) Taking bearings with simple transit; (B) Construction details for simple transit; (C) Data sheet of bearings and distances; (D) Finished map of playground area.*

object. Follow the same directions for using a simple transit, but replace the circular protractor with a compass on a flat surface. Here are the potential problems you have to deal with when using a compass.

1. I find it very difficult to deal with the concept of declination with children (or with myself for that matter). *Declination* is the difference between true north and magnetic north wherever you are using a compass. It is different in every location and even changes a little bit in each location from year to year. I find it much easier at this level to ignore declination and just orient maps and measurements to magnetic north rather than true north.

   To find bearings to the objects on your boundary, set the compass on the table, align the floating compass needle with the north bearing arrow on the housing of the compass, and find a boundary object that is located at magnetic north. Have this object be your first, fixed boundary object. Then proceed to take bearings of your other boundary objects from here. Again, refer to compass instructional materials for information on how to teach compass use.

2. Be aware of the presence of magnetic fields that may deflect your compass both inside and outside. I was mapping a tiny city park one year with a group of teachers and we kept getting screwy data. We finally realized that the electrical power supply for a large nearby building traveled up a utility pole located at one of the corners of the park. The electricity flowing through these lines created a strong magnetic field around the wires that interfered with our measurements. Similarly, any electrical lines or metal on your playground will deflect compasses. Therefore, don't place your compass support table over an iron storm drain or next to the swing set.

   Additionally, it's very difficult to use compasses accurately inside. Most student desks have some iron components in them and there's lots of circuitry in the walls. Thus, to translate information from outside to inside, you should use a protractor to measure angles rather than the compass you used outside.

3. The flip side of all of these cautions is that compasses connote adventure, mystery, and exploration, and some children are intrigued to use real scientific equipment. If you're willing to invest the time in teaching compass use through extensive games, orienteering, and other real-life applications, then using compasses for mapmaking will be appropriate and exciting. After all, what explorer would head out into the unknown without a compass (or a handheld global positioning system instrument)? Learning to use a compass is part of the *learning to survive on your own in the woods* theme that is strongly emergent at this age. If you've got a group of children intent on learning about fire building, wild edible foods, and survival skills, then take a chance with teaching compass use.

It's inevitable that the question of representing contour intervals and height is going to emerge if you're doing any kind of environmental work with children at this age. Representing topography presents a whole new level of conceptual challenge for children. I caution you not to approach this lightly or with children who are too young. I remember the challenge of learning to interpret contour maps in my college years. Even then, I found it to be tough sledding. Most adults don't understand them. And while you can get children to grasp certain simple patterns (e.g., wherever there's a set of nested circles is a hilltop; lots of lines scrunched together show a cliff), getting them to navigate with topographical maps or to make contour maps of a hilly area near the school is a different story.

Following the small-world principle, it's best to start contour mapping with a desktop activity. So let's get small again.

Mapping Miniature Mountains

These activities work best with exactly the right materials. Of course, you'll have to adapt the list to your circumstances, but it's worth the energy assembling the following materials. For both of these methods you'll need:

> A lot of modeling clay (plasticine, not a water-based clay)
> Large plastic basins (broad and deep transparent ones are best)
> Sheets of Plexiglas or clear plastic that fit over basins
> One-eighth-inch-diameter wooden dowels (cut in pencil-length sections)
> Trays or shallow cardboard boxes
> Clear acetate sheets
> Acetate markers
> Toothpicks
> String
> Rulers
> Scissors
> Sand

The plasticine and water method

Have small groups of children create miniature mountains out of plasticine that will easily fit inside the plastic basins you have available. The mountains should be about five inches high. If you think some of the groups are up to the challenge, encourage them to make mountains with interesting features—twin peaks, a cliff, a lake, some steep sides, and some gradual sides. Once the mountains are complete, follow these steps.

1. Place the mountains inside the clear plastic basins. Center them so that children can access all sides of them.
2. Tape a small plastic ruler to the outside of the basin, or mark a piece of masking tape at half-inch intervals.

3. Pour water into the basin so it fills up to the half-inch level. The mountain has become an island surrounded by a lake. Since water lies perfectly flat, parallel to the surface of the earth, it intersects the mountain at exactly the same height all the way around.

4. Use a sharpened pencil to trace around the shore of the island at the water's edge.

5. Add more water to the lake to a depth of one inch, and trace around the new shore of the island at the water's edge.

6. Continue adding water and tracing around the water's edge until the water covers the plasticine. Then empty the water out of the basin.

7. Place the Plexiglas or clear plastic over the basins and secure them so they don't slide around. Tape the acetate sheet on top of the Plexiglas.

8. Looking down on the mountain from a consistent point, trace the lines drawn around the mountain. You should get a set of nested circles with a little bit of space between each line.

Once the tracing is complete you have a contour map of the mountain. Have the children look at the maps and try to see patterns (Figure 6–6). Ask the children these questions:

How can you tell where the top of the mountain is?
How are steep slopes and gentle slopes shown on the map?
How can you tell whose mountain in the class is the highest by looking at the maps?
What would the map look like if the mountain were a volcano and there was a round crater in the top of the mountain?

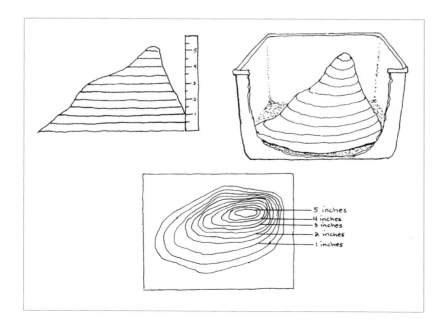

FIGURE 6–6
Creating a contour map of a miniature plasticine mountain. Water is poured into a plastic basin and contour lines are drawn on the mountain at different elevations.

The sand and string method

If you don't want to face the water hazard of the previous activity or you can't find clear plastic basins, this approach works almost as well. However, remember how much custodians dislike sand. Using slightly moistened sand, create sand landscapes on the trays. Though it's tempting to make the landscapes complex, encourage the students to keep it simple. Two mountains are enough. An island isolated in the center of the tray simplifies matters.

1. Mark the dowels at one-inch intervals to make depth-measuring probes. Coloring in every other inch makes the probe easier to read.
2. Have children use these probes to make holes at the spots where the sand is one inch deep all around the mountain. Slip a toothpick into each hole.
3. Connect together all of the one-inch locations with a piece of string—a kind of dot-to-dot exercise.
4. Use the probes to find the two-inch depth, and connect these holes together with another piece of string. If children have two mountains, at some point they will have two ovals of string rather than just one. You will have to demonstrate the process of laying out the contour lines. (This is similar to the process of creating depth intervals in the "Puddle Mapping" exercise.)
5. Continue until the highest elevations are delineated with string.
6. You can transfer the string contours to a flat surface by placing a piece of Plexiglas over the model and tracing onto acetate.
7. Another interesting alternative is to take "aerial photographs" of the mountains at this point, thereby transferring the three-dimensional image to two dimensions. Then the challenge becomes matching the photographs to the appropriate landscapes.

Both of these small-world contour mapping exercises rough in the concept of representing differences in heights on maps. Once these activities are complete, you might try going in the opposite direction—giving the children simple contour maps and having them create these landscapes in sand. This will allow you to see if they have really grasped the concept. The next step is to increase the scope and move outside.

Tackling Topography

Choosing the area to map

Put some time into choosing the areas to be mapped. There's a lot to think about. Here are some guidelines:

- Try to find an area with some substantial changes in elevation. A hill of twenty or thirty feet will really make the process a lot more tangible. An elevation difference of two or three feet doesn't have much drama to it.

- On the other hand, keep the area small. One hundred feet square is quite large enough. Making too big a jump too quickly will undermine your efforts.
- If possible, choose a meadowy area rather than a forested area. Negotiating around obstacles will distract the students. On the other hand, stone walls, playground structures, or a few bushes and trees will provide good landmarks for a contour map.
- Choose an area that has a usable baseline for one border—the side of a straight road, a fence, the edge of the asphalt playground.
- Optimally, find a discrete, small hill that is similar to the plasticine or sand hills you have mapped inside. For a teacher workshop, I once used three eight-foot-high piles of gravel on a nearby construction site. These provided the perfect next-step challenge for the budding cartographers. Most likely, however, discrete hills like these will not be available, and you will have to settle for a slope. The problem with a slope is that the contour map will not wind up looking like a set of nested ovals as in your indoor maps. In this case, it might be useful to create a rough model of the area you're going to map and suggest the general configuration of the contour map you'll be getting before starting outside.
- Choose an area where you can drive stakes into the ground and leave them there for a while, feeling relatively sure that they won't be disturbed and that people won't trip on them. It's unlikely you'll be able to finish this project in a day.
- Create a base map to work from to put the topography onto. You have a couple of choices. You can have the students create a base map using the baseline and offset or plane-tabling techniques described previously. You can also enlarge a local street map that includes the area or use the architectural site plan for the school.

Once you have chosen your area and have a suitable base map (which you now see is a job-and-a-half in itself), you'll either have to find or make the appropriate equipment.

Making an alidade and rod

The core problem in contour mapping is figuring out how to tell the difference in elevation between one point and another point. Here's an easy way to think about solving the problem. Stand at the bottom of a flight of stairs. Look straight ahead of you, holding your head as flat as possible, and find the step that is at the same level as your gaze. Let's assume that it is approximately five feet from the floor to the level of your eyes. The step that you are looking at is on a line with your eyes that is parallel to the floor, so that step is at an elevation five feet above the level of the floor. Now go up and stand on the step you were looking at and find the next step that is level with your gaze. That second step is now ten feet above the elevation of the floor. Finally, stand on the second step and see that the top step comes up to the level of your waist. Figuring that the distance from the bottom of your feet to your waist is three feet, we now know that the second floor is thirteen feet above the elevation of the first floor.

Outside, you will be doing the same process, but since hills usually aren't as steep as steps, you have to look over longer distances. And since there aren't a bunch of notches on the side of the hill like steps, it's harder to find exactly the point you think is level with your eyes. To extend our senses and make the process accurate, we use an *alidade* and a *range rod*. An alidade is a tool that allows you to level your gaze across an expanse of space. A range rod is a long ruler that allows you to measure the difference in height between two locations. Together, these tools give you the means to collect topographical data.

Construct one alidade and range rod for each team of four students (see Figure 6–7). Begin by nailing inexpensive wooden carpenter's levels to a four-foot board that measures approximately one inch by two inches. Make sure that the carpenter's level is joined to the board at a right angle. Attach a screw eye to one top surface of the level to sight through, and a finishing nail at the other end of the level to provide a sighting line. Attach a three-foot piece of string to the bottom of the level, using another screw eye, and tie a large washer to the bottom of the string to create a plumb line. Cut the bottom of the board so it is pointed so the distance between the sighting line and the bottom tip of the board is exactly four feet.

Construct the range rods by cutting six-foot lengths of the same one-by-two-inch boards. Mark off one-foot lengths on the rods and then color alternate feet a dark color using paint or permanent markers. The alternate shading of feet gives the surveyor easy points to focus on when sighting over long distances.

It will be important to practice with the alidade and range rod in the classroom before using them in the field. Children will need practice holding the alidade level by keeping the plumb bob (the string and washer) vertical to the edge of the upright, keeping the bubble centered in the level, and reading height off the range rod. They should also practice measurements.

You'll also need a bunch of stakes to drive into the ground at the points you've measured. Make about thirty or forty of these and paint some red, orange, green, blue, and black. You will use these to signify points with similar elevations. Give several markers to each team.

Finally, each team will need a copy of the base map on which to record the field information, and a long tape measure. Now you're ready to head out into the field. To recap, the materials you'll need for contour mapping are:

- an alidade and range rod for each team of four students;
- a base map for data collection for each team;

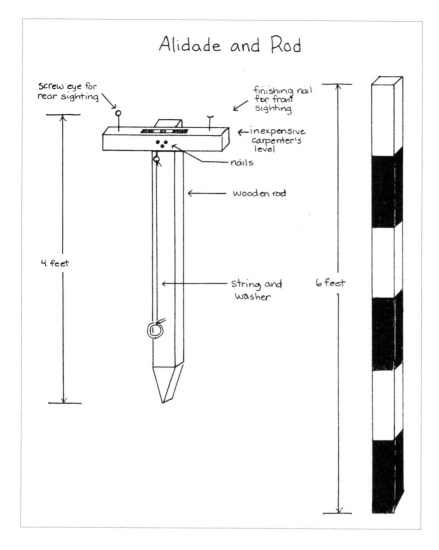

FIGURE 6–7
Alidade and rod construction details.

- elevation markers for each team; and
- a tape measure for each team.

Collecting field data

Let's assume you'll be measuring a hillside slope on the edge of the school playground. We'll also assume that the baseline you are using is the flat edge of a road or the wall of the school and that this baseline is fairly level to the bottom of the hill that you'll be measuring. Here's how to proceed.

1. If it's a long way from your baseline to the bottom of the hill, establish a second parallel baseline at the bottom of the hill. Do this by using the baseline and offset technique. It will be easiest if the ele-

vation of this second baseline is the same as your original one. If not, you'll have to measure the elevation difference between the two. If your baseline itself is not level, find the lowest point and then measure the difference in elevation between different points of the baseline.

2. Establish four points along the baseline you're working from. Let's say your baseline is sixty feet long. You'll have a point at each end and two other points that are twenty feet apart. From these points, find points at the top of the slope that are along lines that are perpendicular to the baseline. You will be measuring the elevation differences along these lines. Drive stakes into the ground at these points both on the baseline and at the top of the hill.

3. Starting at a point on the baseline, one pair of students uses the alidade while another pair of students uses the range rod. While one student sights through the alidade, the other student makes sure that the alidade is kept level. The object is to place a stake at sequential points along the line that are each two feet higher than the previous point. The alidade students direct the range rod students to move along the line until an agreed upon elevation difference (of two feet). When the alidade students can see the two-foot mark on the range rod (measuring up from the bottom of the rod), the ground level at the bottom of the range rod is two feet above the point at the bottom of the alidade (see Figure 6–8). The range rod team drives a stake into the ground at that point.

4. The alidade team now moves to the point just marked while the range rod team moves farther up the hill. From this new point, the alidade team instructs the range rod team to stop at the point that is again two feet higher than this new point. (This is analagous to walking up the steps.) This new point where the range rod team is located is now four feet (two feet plus two feet) above the starting point on the baseline.

5. The student teams continue in this fashion up to the marker at the top of the hill. If other teams have been working in a similar fashion along the other lines from the baseline, then you will now have a set of markers indicating two-foot, four-foot, six-foot, eight-foot, and ten-foot contours up to the top of the hill.

6. After you have completed marking all of the elevation points, you need to find the distance between each of the points. I think it's easier to do this as a separate task rather than in conjunction with marking the elevations. The correct way to measure distances is horizontally, along the level line between the two elevation points rather than along the ground. This is done by holding one end of the tape at the base of the range rod and extending it until it meets the midpoint (two feet above the ground) on the alidade. The actual difference in distances between measuring along the ground

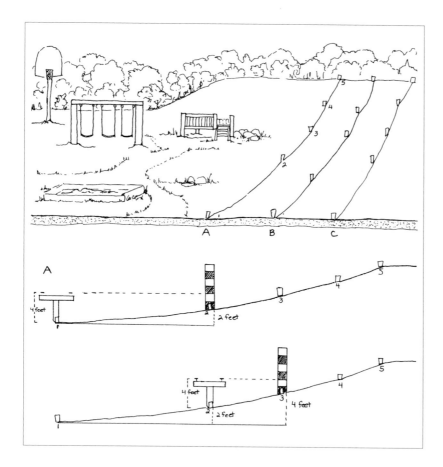

FIGURE 6–8
*Using an alidade and rod
to collect elevation
information for contour
mapping.*

and measuring horizontally will be fairly small, so it may not be necessary to introduce this subtle distinction. You can make this decision based on the sophistication of the students you are working with on this project.

If you have a hill with a small distinct top that slopes down on all sides around it, you can start your topographical measurements from one point at the top. From here, you can take elevation measurements along direction lines plotted with the use of a compass. Lay out the sighting lines to the N, NE, E, SE, S, SW, W, and NW. The alidade team then directs the range rod team to descend along one of the lines until they get to a point two feet below the level of the top of the hill. To find this point, the alidade sighter will have to be looking at the 0 mark (the top) of the range rod. (Six feet minus four feet—the height of the alidade—equals two feet.) The range rod team marks this point, the alidade team moves to this point, and so on to the bottom of the hill.

Similarly, this technique of using angled sighting lines rather than right-angle lines can work from the bottom of the hill as well. The alidade team can work from one or two different points on the baseline and measure elevation differences along any line of their choosing. This simply requires that the teacher and students understand the use of a compass and can take the bearing of the sighting lines.

Drawing the contour map

Drawing the contour map is fairly straightforward after all of the hard work in the field. Determine a scale, transfer your base map to a large sheet of paper, and then locate your baseline on the map. Next, locate the starting alidade locations along your baseline.

Creating the map will be easiest if you lightly draw in the sighting lines that you used for determining elevations. If you used sighting lines that were perpendicular to the baseline, these will be easy to draw in. If you used sighting lines determined by a compass bearing, you will need to use a large protractor to determine the correct angles for the sighting lines.

Once the sighting lines are sketched in, find and mark the location of each of your range rod points. These are the points where you drove marking stakes into the ground. Mark each of these locations with the correct elevation above the baseline.

Now, connect together the similar elevation points with a contour line, and identify each as the two-foot contour, the four-foot contour, and so on. On many topographical maps, the line indicating a multiple of ten or twenty feet is bolder, and you may want to do this as well. Make sure that none of the contour lines cross each other. Locate all of the other features—stones, drains, playground equipment, trees or shrubs, and so on—on the map as well. Add a title, scale, legend, compass rose, and you will have a completed map.

Once you've completed the map, it will be valuable to ask questions that require your students to interpret and understand it:

> Where are the steepest places on the hill? How can you find them on the map?
> What is the elevation on the top of the hill?
> What are the elevations of some of the objects that are located in between two contour intervals? How do you know?
> If you could make the paper big enough, where would the unconnected contour intervals of the map actually loop back together and connect?

An appropriate extension of this activity is to re-create the actual shape of the area you mapped by building up the contour intervals on your map with cardboard or foam core sheets cut to the shape of each contour interval. Going from three dimensions to two dimensions and back creates a deep understanding of the relationship between the real place and the map. Having this experience with a small piece of the world will allow students to enter into maps of much larger areas and reconstruct the landscape in their mind's eye.

Craig Altobell, Fifth- and Sixth-Grade Science Teacher
Cogswell Memorial Middle School, Henniker, New Hampshire

Many teachers like to arrive at school forty-five minutes early so they have time to set up the classroom, review their plans for the day, photocopy worksheets, and have a cup of coffee with a colleague before the buses arrive and the swirl of activity begins. During the spring of 1996, Craig Altobell tried a different approach. Arriving at school by 7:00 A.M., he extracted his boat from the high grass at the edge of the meadow and slipped it into the waters of the Contoocook River, which flows along the edge of the school grounds. Then he kayaked upstream for about twenty minutes, turned around, and drifted downstream for another fifteen. He still had enough time to pull things together before his homeroom students arrived at 8:00.

One of the mapmaking activities during the year that really intrigued me was a unique marriage of abstract curriculum goals and local geography. When he was not studying the river, Craig did a variety of other district-mandated science units, including paper towel testing, bubbles, studies of volume and density, and, of course, the solar system. Despite all that model making that had happened in the earlier grades, the upper elementary students were still baffled by the distances involved in the solar system. Ninety-three million miles—is that as far as from here to Chicago?

In response to this confusion, Craig came up with an ingenious solution that really worked for his sixth graders. His approach was based on a curriculum unit entitled "Earth as Peppercorn." Assuming that the sun was an eight-inch ball, the planets were analogized to objects that were proportionately in scale with the sun. Mercury and Mars were about the size of a pinhead, the Earth and Venus were peppercorns, Jupiter was a

Craig and his students were in the midst of a yearlong theme study of the Contoocook River watershed, and this early morning landscape immersion served as both meditation and productive curriculum planning time. He said, "I felt connected to the river, the organism we were trying to get a sense of, and I got new ideas about activities to do with the children. My encounters with mink, ospreys, beavers, and herons were the raw material for the stories that initiated our activities for the day." The yearlong study had taken the students up to the top of Mount Monadnock to get a long view of the upper watershed of the river, down the river in canoes, and into the river to study macroinvertebrates and do water testing. Science was balanced with art in a continual emphasis on writing poetry and creating exquisite collages based on experiences in the landscape. Craig wanted his students to develop both love for and knowledge about the watershed that supported their daily lives.

pecan, Saturn was a hazelnut, and so on. Craig reviewed the mnemonic device for remembering the order of the planets: My Very Enormous Mother Just Sat Upon Nine Pins. Then he posed a problem.

He placed the sun on the floor near the front of the classroom and said, "Using the same scale as the relationship between this ball and the real sun, place all of the planets at their correct distance from the sun." Many of the students correctly showed that Mercury, Venus, Earth, and Mars are closely clustered together and that the other planets are farther out, but none of them grasped how far away the planets really are from each other. They showed the inner planets about eight or ten feet from the front of the classroom, and Pluto wound up toward the back of the room. In fact, at the same scale as Earth as peppercorn, it turns out that the distance between the sun and the Earth is twenty-six meters, at least a couple of classrooms down the hall, and the distance to Pluto is a mile!

The next order of business was a field trip to the neighborhood next to the school. Craig prepared the students for the field trip with a worksheet that asked them to calculate the appropriate distances between each planet based on a number of paces. Craig had the students practice pacing so they could adjust their stride to make two steps (one full pace) equal one meter. This meter or pace equated to 3,600,000 miles in the sky. Once these calculations were complete, the class went outside. The sun was placed on the sidewalk, next to a telephone pole by the front door of the school. Mercury was ten paces away, by the stump of an old tree. Venus was nine paces far-

FIGURE 6–9   *A section of sixth grader Russell Durgin's map of the Henniker, New Hampshire Solar System Walk.*

ther, by a crack in the sidewalk. Earth was seven more paces, barely at the edge of the school building. It was only fourteen more paces to Mars, in front of the library next door. Then it was a stretch, ninety-five paces, to place Jupiter's pecan at the traffic island in the middle of town.

Now the reality of space started to take shape. It was 112 paces to Saturn, at the edge of the downtown buildings of Henniker, and then 249 paces to Uranus. By now the kids were way out the Old Concord Road, out in farm country along the river and a long way from the school and the village. It was a tiring 281 paces to Neptune out by the cemetery (by the Pillsbury headstone to be exact), and 242 paces to Pluto's pinhead a mile from school, at the bridge across Amey Brook, a tributary of the Contoocook.

Back in the classroom Craig gave the students the assignment that would keep them busy for the next three science periods:

Henniker Solar System Map Walk

Create a map of our solar system walk that accurately shows the landscape and town landmarks along our route and places the planets in their correct locations. Your maps need to have a scale, be readable and easy to understand, have a title, show the number of paces between each planet, and be done neatly. Try to portray our route through Henniker as clearly as possible.

The remarkable maps created by the students are a sight to behold (see Figure 6–9). Most students chose a scale of 1 centimeter = 10 paces,

FIGURE 6–9 *Continued*

which made their maps four feet long. In panoramic splendor they showed architectural details of downtown buildings, the ATM machine, a dress displayed in the clothing store's window, stone walls, tractors in the fields along the river, evergreens bordering the road. One student recruited her father to help drive her around and take photographs of all of the town buildings and locations of the planets. Her photographs, dated April 3, replete with icy roads and bulky snow piles, documented the unending winter of 1996. They were fitted into a drawn landscape, and at a scale of 1 centimeter = 5 paces, they created an eight-foot-long map of both her class and family exploration.

This project had an on-target quality to it that is unusual in the abstract study of "long ago and far away" topics in the elementary curriculum. The core element that made it work was the panoramic map of the walk through town. By mapping the solar system (the unknown) onto a map of the town (the known), the students were able to kinesthetically and conceptually grasp how big the solar system is and how small the earth is in relation to the solar system. When they hear about a space probe traveling to Jupiter, they can appreciate how gigantic a distance that is compared to sending people to the moon or exploring Mars. Craig reflected, "When this was done, many of the students had a real sense of the vastness and emptiness of space. It makes the wonder of life on earth and life in our own backyard a bit more significant."

The true test of understanding came two weeks later when Craig talked to his students about a meteorite, capable of inflicting vast damage on impact, that passed within two hundred thousand miles of earth. This event had no significance for most children because they didn't have a way of grasping what this distance means. But Craig's students replied, "That's only the distance between the moon and the earth. Yow! That's close." At the scale of their walk, two hundred thousand miles is only five centimeters, a fraction of a pace, barely the width of a child's hand on the sidewalk in front of the school. In terms of the size of the solar system, that's a pretty close call.

Craig's approach works because his curriculum is based on knowing the place where his students live. With a firm rooting in their home turf, they are able to reach up and out to start to understand the planets and perhaps the stars. In our aspiration to give children both roots and wings, we need to remember that it takes a long time for roots to develop before they will support the weight of lofty abstraction. Just as we make sense of tools by analogizing them to our body, we can make sense of large geographies by analogizing them to local places: A door hinge works just like my elbow, this trickle coming from the melting snowpile is just like the meltwater river pouring from the glacier. As above, so below. Let's remind ourselves that the below is right outside the classroom door.

7 | ENTERING THE LANDSCAPES OF CHILDREN'S LITERATURE

THE INNER WORLD

The Friday afternoon sun streams through the window of the playroom into my father's New York City apartment. Since my parents' divorce when I was three years old, a couple of years ago, I spend every other week in the city with my father. We dine out at Schrafft's, build models, and feed the pigeons, and when we visit my aunt on Long Island, my father makes treasure hunts for me and my cousin. My father reads to me a lot, which I really like because my mother never seems to have time to do it. He always finds good books, and right now I notice a book I have never seen, spotlighted by the sunlight, on top of the toy box. I can't read the title but I open it up, and what's this? There's a map on the inside cover where usually there's just plain paper. This map must show all of the places where the story happens. I have never seen a book with a map in it before—didn't realize that stories could have maps—but I know I'm going to love this story. I can't wait for him to start reading it to me. It's not bedtime but I call, "Daddy, could you start reading this right now!?"

This memory of discovering the map in Ruth Stiles Gannett's *My Father's Dragon* is one of the vivid memories of my early childhood (see Figure 7–1). I have no recollection of chasing our pet beagle around the house with a carving knife when I was four years old, but something about the map and the imaginary world it promised captured me even then. As my father read the story, we flipped back to the endpaper to trace Elmer's progress in his quest to release the baby dragon from the clutches of the savage animals who kept her enslaved on Wild Island. Tiny pictures of the fierce animals that Elmer had to outsmart along the path made it very easy for me to understand how all of the parts of the story fit together. The other books in this trilogy, *Elmer and the Dragon* and *The Dragons of Blueland,* also contain maps. I reread them often in childhood and have read them more than once with my own children. The map allows us all to enter into the story, to get one step closer to really wandering about on Wild

Island ourselves. We feel the jungle pressing in on Elmer as he tiptoes through the dense foliage.

Along with many other young readers, I was entranced with the map in Robert Louis Stevenson's *Treasure Island*. If there was a map, then this had to be a true story. Is this what treasure maps really looked like? I started searching for books with maps—somehow these stories always seemed more exciting. The map in Joy Adamson's *Born Free* and the story of Elsa the lion brought me to the African savannah, and Sir Edmund Hillary's account of climbing Everest in *High in the Thin, Cold Air* had one of those detailed expedition maps with dates, elevations, and annotated events. Now, I often read with an atlas propped open next to me so I can follow the narrative of geographically based books that don't have maps. At some point I began to wonder, why don't all books have maps, and if they don't, why not make them?

This question unavoidably led me into making maps of stories for and with children and helping them make their own maps. Just as mapmaking helps us to know and love our neighborhoods and local watersheds, making maps of stories helps us to live inside of the story space in a deeper and more meaningful fashion. Integrated literature-based curriculum serves to

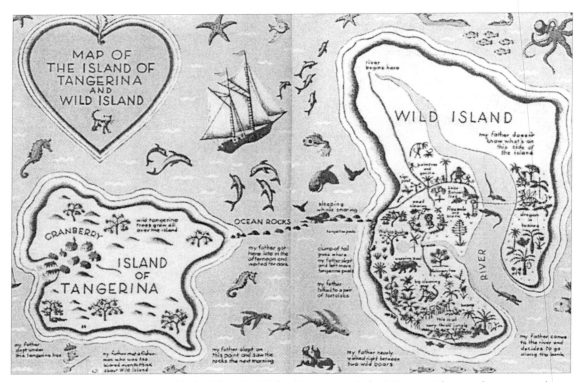

FIGURE 7–1   *Endpaper map of Wild Island from Ruth Stiles Gannett's* My Father's Dragon. *This map for young readers has a highly pictorial quality.*

counteract the superficiality of the television mind-set. On TV, one story is quickly submerged by another and another. At the end of an evening, you can't even recall the first story you watched. Moreover, the stories often unfold in a denatured interior studio that has no geography, no sense of place. By entering into stories and living inside them for a while, we cultivate patience and appreciation. Children see that stories take place in landscapes and that people are dependent on the natural and built environments for their health and livelihood.

Whether stories are about real events or are fictional, they create geographic worlds that can be visualized with the assistance of a map. And it is just as important to help children map the worlds inside their heads as it is to map the outside worlds. Composing an integrated landscape from the bits of stories and scenes in a book is a perfect example of conceptual synthesis. Intelligence is often described as the ability to see the relationship between the parts and the whole. Visionaries are able to envision a reality that does not exist and then take the steps necessary to shape this new reality. Helping children create maps of imaginary places is a pure example of constructivist pedagogy.

The story map that the child creates is analogous to the concept sketch of the architect. The architect hears that the client wants a sunny living room, two bedrooms, a yard safe for young children to play in, and a porch facing the street so she can talk with her neighbors. The architect looks at the lay of the land, the trees, the sources of noise, and the location of the utility lines. Then she assembles an imagined image, a map, of how all of these pieces fit together in a functional and pleasing manner. Through cultivating fantasy mapmaking with children we are developing the graphicacy skills that engineers, landscape planners, clothing designers, graphic artists, novelists, and teachers need every day. As inside, so outside.

DIFFERENT STROKES FOR DIFFERENT FOLKS

A wide range of possibilities is available to teachers who are interested in enhancing the classroom literature experience with maps. Simply choosing books with maps is a first step. Christopher Robin's map of the Hundred Acre Wood in *Winnie-the-Pooh* by A. A. Milne makes a great introduction for early readers. But since this map is too small to be seen by twenty anxious listeners, it's valuable to make an enlarged version of it. This can be done by successive photocopier enlargements, by a hand-drawn enlargement done by the teacher, or by an enlargement made by projecting a transparency of the map onto paper taped to the wall and then tracing the map onto the paper. This last option can be done by a small group of careful children. Most maps in children's books are highly pictorial or panoramic because children's book illustrators intuitively understand that this is the kind of map that young children easily grasp.

Books for older readers have more abstract maps. The map in *The Ghost of Lost Island* by Liza Ketchum Murrow, a book for upper elementary students, is drawn from a high panoramic perspective with a smaller scale aerial view inset. It's quite different from the Wild Island map targeted for young readers. This more sophisticated view fits perfectly with the developmental capacities of older children. The realism of the map appeals to the older reader because it makes the island feel like a tangible destination. And the map provides enough pictorial detail to allow the reader to follow the narrative through the landscape (see Figure 7–2).

Though most children's books do not have maps on the endpapers, be on the lookout for books that have full-color, panoramic landscape views as part of the text. These illustrations are often just as good as, if not better than, maps for helping children get a sense of the whole story. They are, however, more difficult to enlarge for whole-group use. *The Fox Went Out on a Chilly Night* by Peter Spier (1961) has a succession of panoramic views

FIGURE 7–2 *Endpaper map of Lost Island from Liza Ketchum Murrow's* The Ghost of Lost Island. *This map for older readers uses a high panoramic, almost aerial perspective.*

that allows the reader to follow the fox's journey from its den to the farmer's pen and back again (Figure 7–3). *Rabbit Island* by Jorg Steiner (1977), with illustrations by Jorg Muller, has an extravagant centerfold containing a panoramic view of the story landscape.

But alas, most books don't have maps. While this is an obstacle for us map-loving readers, it is an opportunity for adventurous classroom teachers willing to take things into their own hands or to put things into their students' hands. First, recognize that some books lend themselves to mapping while others don't. Choose books that involve some movement through a setting or a landscape. *Little Red Riding Hood* is a perfectly mappable story. Many of the Frances books by Russell and Lillian Hoban, with an emphasis on personal relationships, are less suitable for mapping. I particularly like to choose books in which the illustrations provide a set of clues for how the parts of the story fit together. John Schoenherr's illustrations in Jane Yolen's *Owl Moon* provide a set of sequential and overlapping views of the moonlit woods at night that lend themselves to the creation of a panoramic view.

Depending on the teacher's graphic confidence, the age of the children, and the size of the class group, there are three literature and map strategies available for teachers.

FIGURE 7–3   *Teacher-made map based on Peter Spier's* The Fox Went Out on a Chilly Night.

Teacher-Made Maps

Teacher-made maps are appropriate for students of all ages. Some teachers like to make maps of the story prior to reading it to the children, while others like to create a map incrementally, chapter by chapter, as they read to the children. Similar to the beautiful chalkboard images that Waldorf teachers create of the stories they tell to their students, teacher-made maps help to weave a spell that brings the story alive. And don't think this works only with young readers. I know a sixth-grade teacher who has made exquisite maps for the Lloyd Alexander trilogy books that he reads to his students.

Modeling your own artistic and graphic experimentation plays a crucial role in encouraging students to practice their graphic skills. If the students see you working to develop as an artist and mapmaker, they will strive in a similar fashion.

Teacher-and-Student-Made Maps

Teacher-and-student-made maps involve a collaborative effort between you and your class. This process is guided by the same principles as some of the mapmaking projects described in earlier chapters. Just as it was necessary for the teacher to provide a prepared outline map to initiate a playground mapping exercise, it often is necessary for the teacher to provide the initial scaffolding, or skeleton, for a story-mapping experience. The idea is to create the map of the story together as it unfolds. For example, if we start with Little Red Riding Hood's house on one side of the map and Grandmother's house on this side of the map, where should the woods start? Where does the wolf meet Red? Where does Red pick flowers? What path does the wolf follow to Grandmother's house, and which path does Red follow? As the story progresses, new places are added to the map.

For this process, it's important that the initial map is a draft with moveable parts that can be adjusted as necessary. If the map is done on the floor, blocks and sketches can be moved about to fit the story elements. If it's done on a wall, a magnetized surface can facilitate moving drawn elements around. I have used Post-It notes for this part of the process, but their size is sometimes a limitation. Once the story is finished, the map can be cast in ink, but during the development it is best to think of the map as a work in progress.

Student-Made Maps

Student-made maps will be just the ticket for a select group of students in your class. In the upper elementary grades particularly, some children are ineluctably drawn to maps and will jump at the opportunity to make their own maps of imagined places. As an alternative to having children write a book report, provide the option of having them make a map of the book they read. Provide a set of parameters for what the map needs to include, and let them have at it. For example, you might require that they include the title, the author, at least ten story locations, some indication of the path that the characters follow through the story, a compass rose, and an approximate scale. Help beginning mapmakers by providing examples of books with endpaper maps. When students are faced with a specific illustration problem, such as how to show seaside cliffs in panoramic view, I try to provide them with existing maps with those elements.

A good real-world challenge for students is to make a map for a library book that can be taped into the library copy. This ups the ante aesthetically and can enhance the readability of library books. This process requires clarity of illustration, gets the student to consider the problem of too much "noise" in maps, and presents size constraints that have to be dealt with. It does make sense, though, to have children work on large images and then reduce the image to book size. Nonetheless, they need to start with a shape that is proportionately similar to the shape that the map eventually needs to be to fit into the book.

Of course, these strategies can be mixed and matched. Following are different examples of literature-based mapping at five different grade levels.

CLASSROOM PORTRAITS: LITERATURE-BASED MAPPING PROJECTS

Deanna Avery's First-Grade Classroom Project Nantucket, Massachusetts

Bill Peet's *The Kweeks of Kookatumdee* provides a sample mappable landscape for young readers. Kweeks are large birds with small wings that live on the island of Kookatumdee. The kweeks cannot fly. Their only food is the fruit of the one and only ploppolop tree on the island. Whenever fruit ripens, the kweeks fight over the limited number of ploppolops. Though they try to figure out a way to distribute them evenly, nothing works, and worse, one kweek named Jed starts to dominate. Soon Jed is eating all of the ploppolops and the rest are going hungry. What are they to do? When one of the kweeks is forced off a cliff by Jed, he discovers that he has lost so much weight that he can fly. He teaches the others to fly and, leaving Jed behind, they all fly across the ocean to another island with dozens of ploppolop trees.

This simple story, with simple geography, was Deanna Avery's choice for a set of mapping and writing activities. After reading the story to the class, she had the children recall the landforms from the story. These included both islands, the sole ploppolop tree, the ocean, and a cavern where the kweeks slept. Then she asked the children to create plasticine models of the story in groups of four. The children had already completed some classroom-, playground-, and bedroom-mapping activities, and she wanted to extend mapping into the realm of imagination. She explained,

I wanted them to see that stories can be interacted with, that they can be entertaining but also that there is a great deal of detail and information within the text. I wanted to appeal to those children who like to pull together details and to those who need and desire to work with their hands constructing three-dimensional things. I wanted to provide them with the opportunity to use their imagination and their hands to give form to a story, thereby giving them access and power in the literary world.

Though some of the groups created three-dimensional models of the two islands with trees and kweeks, Deanna was surprised to find that some of the groups made one-dimensional clay pictures that looked just like the pictorial maps of neighborhoods that six-year-olds draw. After the models were complete, Deanna followed with a drawing and writing assignment, asking the children to solve the problem of how the kweeks could share the limited number of ploppolop fruit. The children proposed many good solutions: "The kweeks cood cut them in haf. Wen the Big Bule [Jed] koms saye you haf to sare. Thy lind up so thy cood get ploppalps so thy cood eed."

Next, Deanna returned to mapping, asking the children to draw maps of the story based on the models they had made. Having made the models and written, drawn, and talked about the dilemmas in the story, they were now prepared to take on this somewhat more difficult mapping task. Some children again created pictorial maps, while others created very clear panoramic views of one island with a sole tree and another island with a grove of trees. From an assessment perspective, all of the maps demonstrated that all of the children grasped all of the important story elements.

Another writing assignment followed, asking the children to start where the story ended. "What about Jed?" Deanna inquired. "How does he feel all alone back on the island?" The children's writing suggested a range of moral responses: "The kweeks wre happy beces Jed was not thar." "Jed wasint happy! He wos sad but the kweeks cam bac to tac Jed with them to the uthr ilin."

Deanna's culminating activity was to revisit and amplify the previous mapping projects. The children were each given a ploppolop fruit (a tiny box of raisins wrapped in purple tissue paper) on which they wrote their initials. The children then hid their ploppolops in the classroom or on the playground and drew a map to lead a searcher to the location. Every child created a map with enough furniture, landmarks, and labeling to lead a partner to the ploppolop. The literature-mapping activities were gracefully transformed into mapping nearby places as the children moved from imagination into reality.

What were the keys to success in this project?

- The geography of the book was simple and accessible.
- The teacher artfully sequenced the mapping activities in a developmentally appropriate fashion—from modeling to drawing based on the model to creating a treasure map with a purpose based on the story.

- The teacher integrated mapmaking with drawing, writing, and discussion activities focused on child-relevant issues—bullying and sharing.

David Sobel's Second-Grade Classroom Project
Keene, New Hampshire

I carry my fascinations with me and like to revisit them often. So when I needed a mapping project for a course with teachers, I decided to revisit my old friend *My Father's Dragon*, by Ruth Stiles Gannett. Wouldn't it be wonderful to have a model landscape for the story to unfold on?

I used the raised relief technique described in Chapter 6 to make a tool map model of Wild Island and Tangerina Island. I took the endpaper map, imposed a grid on it, and had each teacher take responsibility for gridding up a section of the map. The final map was then about thirty-six inches by forty-eight inches. We transferred the map to a piece of plywood and glued cardboard contour intervals on top of each other to create the topography of the islands. Then, since I wanted the model to be permanent, we created the surface of the islands using plaster gauze strips used for making body casts. You simply dip the strips in water and then smooth them into place on the landscape, much like papier-mâché but with a more finished surface. Painted, it had just the right texture and look of tropical islands viewed from an airplane (see Figure 7–4).

In *My Father's Dragon*, the main character, Elmer, meets a cat who tells him the tragic tale of a baby dragon enslaved by wild animals. Elmer commits himself to saving the dragon, so he stows away on a ship bound for Cranberry on the Island of Tangerina, the closest port to Wild Island. He travels along the coast of Tangerina to a row of rocks that stretch across the ocean to Wild Island. Under the cover of darkness, he crosses the rocks, only realizing later that one of the rocks was a whale. Then he hides in the jungle. Since two wild boars guard the island, he always needs to move stealthily to avoid detection. On Wild Island he encounters tortoises, seven tigers, a rhinoceros, a lion, a big gorilla, and finally some alligators before he can find and free the dragon. For each encounter he has to outsmart the wild animals, who are determined to eat him or detain him in some way.

To make the story come alive, we collected a legion of tiny props to illustrate each event along the way. This became an exercise in scale; once we decided on the size of Elmer and the dragon, everything else roughly had to fit the same scale. We made a boat for him to stow away on, the tiny houses and docks of Cranberry, and the tangerine trees of Tangerina. We glued a set of smooth sea rocks between the two islands, leaving a space for a miniature whale. Then, by scavenging our children's toys and downtown hobby shops, we came up with all of the animals of Wild Island. The project had the same fascination as creating the landscapes for model train layouts. Because all of the pieces were small, they fit into the compartments of a sampler box for chocolates. Opening the box felt like coming upon your grandmother's secret jewelry collection.

Today, second-grade teachers borrow the model and figures to use when they read *My Father's Dragon* to their class. At the end of each chapter, they delicately open the box, take out only the current pieces, and let the children figure out where the pieces should go. It is, of course, a privilege to get to place the pieces, but lots of discussion ensues about exactly the right location. Teachers sometimes go back and read passages from the text or refer the children to the endpaper map to help with the decision. The model resides in an accessible place in the classroom so that the children can use it like a tool map, playing out the story in their free time. Some teachers also use the model to have the children retell the story before they read a new chapter. Having moveable pieces and the miniature landscape helps children to grasp details and sequence in this somewhat complicated story.

FIGURE 7–4
Raised relief map created with plaster gauze based on endpaper map from My Father's Dragon. *A set of tiny figures of the animals and characters allows the teacher and students to re-create the story on the map as they read the book.*

With older children, this whole process could certainly be duplicated. Perhaps this would be a wonderful tool to use in buddy reading programs. The fifth and sixth graders could make models of their favorite books and then use them when they read to first and second graders. If anyone decides to try this out, let me know how it works.

Mary Morrisette's Third-Grade Classroom Project
Westmoreland, New Hampshire

Mary Morrisette, a third-grade teacher in Westmoreland, New Hampshire, used the model-making technique to craft a teacher-and-student-made model of the setting of *Family Under the Bridge*, by Natalie Savage Carlson (1958). This story relates the plight of a homeless old man and a homeless family as they set up residence under a bridge near Notre Dame Cathedral and wander the streets of Paris. She explained that she chose this project because

> a literature map seemed a good way to bring a foreign city and its
> book characters into the classroom. The students had experience
> with geography prior to the Paris story, and we had worked on
> the notion that the world is a neighborhood and that the people
> of the world have traits common to ours.

The map was created on a forty-eight-by-sixty-inch plywood board covered with homemade molding dough and painted to reflect the December climate in France. Mary located the Seine and the Notre Dame Cathedral on the model as landmarks, and then after reading each chapter she and the children made group decisions about where things should go. Buildings were fashioned from boxes and people were designed from small wooden clothespins. Mary used a variety of thoughtful scaffolding techniques during the construction of the model (see Figure 7–5).

To deal with scale, Mary and the students agreed on a qualitative scale. Notre Dame Cathedral would be the highest structure, so all of the buildings had to be shorter than the cathedral. The many bridges in the story had to be big enough so that clothespin people could fit under them, but also not bigger than the cathedral. The smallest character was Jo-jo the dog, so all of the other characters had to be bigger than Jo-jo. To figure out how wide the streets should be in the model, Mary took a small group outside and measured the width of the road and the width of houses. Finding that the road and houses were both about thirty feet wide, they decided that the road on the model should be about as wide as a building.

To orient the children to a foreign place, Mary set up a display of books about France and found a tourist map of Paris—the kind that shows buildings in a pictorial fashion. Mary and the students used this map to locate streets and landmarks for their model, and each child had a copy of this map to use as a reference when working on the model. She also used a transparency to project the map onto the wall during reading sessions so they could find locations as they were mentioned.

FIGURE 7–5
The streets of Paris in this classroom model were based on Family Under the Bridge *by Natalie Savage Carlson. Third-grade classroom, Westmoreland, New Hampshire.*

To organize work on the model, Mary and the students made a list of the characters, buildings, topography (rivers, tunnels, gardens), and landmarks (signs, bridges, statues) as they were mentioned during the reading of each chapter. Then the children signed up for a working group twice a week to make story representations. They chose things from the list that needed to be made and crossed these off once they were completed. As a group, the class decided on the correct locations.

Mary was thrilled at how immersed the children became in the model. She noted,

> During work sessions, the children would bring chairs around the model and work on it at eye level. It was as if they were meandering through the alleyways of Paris when they walked their characters on the model. They didn't just construct the book, they became part of the story. It became a real world for them.

Julie Dolan's Fourth-Grade Classroom Project
Townshend, Vermont

"*Ronia, the Robber's Daughter* [1985] is an awesome book," my fourth-grade daughter exclaimed when she saw it on my desk. Wild and woolly and exploratory, the story unfolds in a lavish wooded landscape that excites the imagination. Author Astrid Lindgren is best known for her plucky heroine Pippi Longstocking, and Ronia is her back-to-nature equal. Ronia is challenged by her father Matt to go into the forest and learn its ways:

She had explored Matt's Fort right up to the parapets. She found her way into all the deserted rooms, where no one ever set foot and she did not lose her way in underground passages, dark pits and cellar vaults. The secret passages of the fortress and the secret paths of the forest—she knew them all now. But it was the forest she loved best, and there she ran free as long as the day lasted.

In her explorations she encounters wild harpies, murktrolls dancing in the moonlight, and other goblinfolk. To outsmart them and accomplish her missions, she needs to learn how to navigate throughout Matt's Wood. She becomes a skilled pathfinder:

It was a long way to the cave, and there was no path for her to follow, but she knew exactly how to reach it. So she walked calmly through the moonlit woods, between the pines and fir trees, over moss and blueberry twigs, past marshland scented with bog myrtle, and past black bottomless pools. She climbed over mossy fallen trees and waded through rippling brooks; straight through the woods she walked, heading unerringly for the Bear's Cave.

Julie Dolan, a fourth-grade teacher at the Townshend Elementary School, realized that the emphasis on path finding in the book made it a fine candidate for mapmaking. Once the class was about halfway through the book, she initiated the mapmaking activities. Julie's class had already completed their study of Vermont geography and had created topographical maps of Vermont. For this assignment, she wanted to challenge some of the students to make two-dimensional panoramic views.

She started by having the children brainstorm a list of significant places in the book. Julie then created a set of simple symbols to stand for the places in the book. She explained, "Creating a map is a pretty abstract process, so I wanted to give them a foundation so they could get started. This provided them with a vocabulary of symbols from which they could start to write the map." With the symbols, Julie sketched out a quick sample map of part of her image of Matt's Forest, showing the location of Matt's Castle, the lake, Greedy Falls, and the trails connecting these features. With this model in mind, the students were then challenged to create their own draft maps using the symbols she had provided.

Similar to using the writing process, conferencing with the children on their first drafts was the next step. Julie assessed the maps by checking to see if landscape features were appropriately connected with each other:

Does the river flow into the lake?
Is Greedy Falls located along the river?
Are Rumphob Holes scattered throughout the forest?
Is the Bear's Cave located far enough away from Borka's Cave?

Students made corrections on the draft maps, and then Julie provided guidelines for the final maps. Using a classic map of Fairyland as a model (showing the integrated landscapes of Snow White, Hansel and Gretel, and many other favorites), Julie required the students to create maps drawn

FIGURE 7–6
*Student-drawn map of
Matt's Forest based on*
Ronia, the Robber's
Daughter *by Astrid
Lindgren. Fourth-grade
classroom, Townshend,
Vermont.*

from a panoramic perspective. The maps had to include symbols or pictures of all of the significant places, and a compass rose. Since Julie had created the "key" ahead of time, the children did not need to include one on their own maps (see Figure 7–6).

The finished maps had a coherent and finished quality to them that was similar to the final drafts of written work that emerge from the writing process. Here are the elements of Julie's step-by-step scaffolding of the process:

1. Brainstorming the places
2. Creating the symbols
3. Modeling her image
4. Having students create a first draft
5. Conferencing and editing the maps
6. Providing guidelines for the final map
7. Displaying a panoramic view model

This well-defined structure eased the children into the process, made each step doable, and made it clear that the creation of a map requires multiple drafts—with an attention to coherence and aesthetics. Mapmaking becomes cartography.

David Sobel's Fifth-Grade Classroom Project
Escazu, Costa Rica

If a book could call out for a map, the voice of *Island of the Blue Dolphins* (O'Dell 1960) would echo off the library wall. This classic account of a Native American girl living alone on an island off the coast of California is a perennial favorite at this grade level. It's the female equivalent of *My Side of the Mountain* (George 1959), and it speaks to girls' interest in personal strength and self-sufficiency.

Karana's band of people have lived on the island for as long as anyone can remember. After a battle with marauding Russian fur hunters, a small group of survivors are barely able to eke out a life on the island. When a sailing vessel from California offers to take the band to live on the mainland, all of the islanders agree to go. But as the boat is leaving, Karana realizes that her young brother, Ramo, has been left behind. She jumps overboard and swims back to the island to take care of him. Ramo, however, is killed by a band of wild dogs, and so Karana is left alone on the island. She builds a fortified home for herself, makes weapons, hunts and gathers seafood, makes clothes from animal skins, discovers caves, and develops a friendship with the leader of a pack of wild dogs. Her adventures take her all over the island, and the dramatic events that unfold invite the reader to visualize the island's landscape.

With a group of better readers in the fifth-grade class at the Blue Valley School in Escazu, Costa Rica, I undertook the creation of a map for *Island of the Blue Dolphins*. The book was used year after year, and the librarian was particularly interested in a map that would enhance the reading-group experience. Our objective was to make a wall display map to be used by reading groups, and smaller, book-sized maps to be inserted into the multiple library copies.

With fifth graders, my goal was to integrate mathematical accuracy, textual analysis, a high panoramic perspective, and a commitment to aesthetics in the creation of the map. To begin, I selected a passage that described the shape of the island, and from this, I challenged the children to devise outline maps that fit the passage from the book:

> Our island is two leagues long and one league wide, and if you were standing on one of the hills that rise in the middle of it, you would think it looked like a fish. Like a dolphin lying on its side, with its tail pointing toward the sunrise, its nose pointing to the sunset, and its fins making reefs and the rocky ledges along the shore.

After determining that a league was approximately three miles, we started to make maps in the shape of a dolphin that measured one league by two leagues. Marianna asked, "What's a dolphin shaped like?" This led us quickly to mammal books and copies of *National Geographic* to give us a suggestion of form. When we had a variety of maps shaped like fat dolphins, skinny dolphins, and jumping dolphins, we discussed which seemed to fit the description best and decided to use that one as the basis for the large map.

At the beginning of the next session, however, one of the children pointed out that the endnotes in the book identified the island as San Nicolas, one of the Channel Islands approximately seventy-five miles west of Santa Barbara. When we found San Nicolas in an atlas, we discovered that our interpretation of "dolphin-shaped" was different from the actual shape of the island. We abandoned our previous attempts and decided to adopt the real shape of the island for our outline map. We adjusted this to the scale of our sketch maps (1 league = 5 inches), and we were ready to begin.

Again, I selected a passage of significant geographical information to get us started:

> The village of Ghalas-at lay east of the hills on a small mesa, near Coral Cove and a good spring. About a half league to the north is another spring, and it was there that the Aleuts put up their tents which were made of skins and were so low to the earth that the men had to crawl into them on their stomachs.

On our original maps we had located the hills that rise in the middle of the island, so we used this as the starting point to locate the places identified in this passage. We found a place for the village by figuring that it had to be east of the hills and about a half league south of Coral Cove since the Aleuts were camped right by the cove. Since we were constantly jiggling the location of places, we identified places on the big map with pictures drawn on Post-It notes, which we adjusted as necessary: "Well, the book says it's one league from the village to the canoe hiding place, but our map shows it's about a league and a half, so we either have to move the village or put the canoe hiding place closer."

Fitting our map to the parameters of the story was a constant challenge, one that was similar to the process of fitting a map to a landscape when you're trying to find your way to a new destination. It taxed our cognitive skills in a challenging but appropriate fashion. I remember the morning when Horacio realized, "Hey, the book says that the Aleut ship is too big to enter Coral Cove, but our ship would slip right in there, so either we have to pinch in the entrance to the cove or make the boat bigger." It was a perfect example of encountering the concept of scale through an authentic, artistic experience.

For the next few sessions, I created a wall chart of places identified in the text that needed to be located on the map. The chart looked like:

p. 1	Coral Cove
p. 4	two rocks that guard Coral Cove
p. 8	cliff overlooking cove
p. 12	sea bass ledge
p. 15	kelp beds on three sides of the island
p. 31	food and water hidden in canoes on south end of island
p. 32	trail to canoe hiding place passing by sand dunes
p. 47	place where Ramo is killed by wild dogs
p. 48	Karana follows wild dogs across two hills and a small valley to a third hill to their den in a cave

The students located these places on their own maps. Then as a group we figured out where they should be on our display map. Once we agreed about locations, children took responsibility for drawing the picture or symbol for different places. First they sketched an idea, and once the students and I agreed that it looked right, they put it permanently onto the map (see Figure 7–7).

Later sessions were a little looser. Each session began with a discussion of new places in the reading for that week, and then children volunteered to put those places on the map. We worked on adjusting scale, but we also agreed that it was OK for Rontu, the leader of the wild dogs, to be

FIGURE 7–7
Students at work on a map based on Island of the Blue Dolphins *by Scott O'Dell. Fifth-grade classroom, Escazu, Costa Rica.*

big with fearsome teeth to show how important he was in the story. The cormorants and sea otters of the island populated our map, and the dry creek beds and towering cliffs started to feel like home. By the end of the project, we had all begun to wander the island in our imaginations. I could feel the seaweed under my feet as Karana hunted for octopus on the tide-exposed sea ledges. I had an image of the layout of her driftwood house on the cliff. This small world had gradually come to life through the vehicle of our cocreated map. Bibliographies of children's books suitable for mapping are listed on pages 163–164.

8 | CHILDREN'S SPECIAL PLACES

Do you remember the special places of your childhood? When you were a young child it might have been in the back of your closet, or under a blanket that you stretched between two beds. Maybe it was beneath the porch steps, or in that alley between the back of the garage and the fence. When you were older, chances are you found a hidden place in the woods behind your house or in the ravine between your neighborhood and the next one over. Maybe a friend had one in the covered-over culvert in the vacant lot down the street or carved out of a corner of a hayloft. Around the world, many children between the ages of eight and eleven are drawn to finding special places for themselves.

The special place is a found or constructed space that children claim as their own, separate and apart from their parents' home. Describing his childhood place, writer Kim Stafford recalls,

> Here was my private version of civilization, my separate hearth. Back Home, there were other versions of this. I would take any refuge from the thoroughfare of plain living—the doll-house, the tree-house, furniture, the tablecloth tent, the attic, the bower in the cedar tree—any platform or den that got me above, under or around the corner from the everyday. There I pledged allegiance to what I knew, as opposed to what was common. My parents' house was a privacy from the street, from the nation, from the rain. But I did not make that house, or find it, or earn it with my own money. It was given to me. My separate hearth had to be invented by me, kindled, sustained, and held secret by my own soul as a rehearsal for departure. (1986)

As Stafford says, finding a special place is a rehearsal for departure. As children move out of the sphere of their parents' influence, they create homes away from home to practice independence. My nine-year-old daughter is constantly talking about living alone in the woods, finding wild foods to eat, constructing a shelter, making weapons to protect

141

herself. It's why Jean Craighead George's *My Side of the Mountain* and Scott O'Dell's *Island of the Blue Dolphins* are archetypical books for fourth and fifth graders. They speak to the longing for self-sufficiency and independence that is part of the developmental move out into the landscape and the community.

A seventh-grade teacher in Keene, New Hampshire, asked her students to write about a personal special place as part of a project on utopias this past year. Of the 105 students in her five English classes, only four said they never had a special place. The other 101 knew exactly what she was talking about and appreciated the opportunity to write about something of importance to them. Recognizing children's deep developmental interests and using them as jumping-off points for curriculum projects is one of the principles of authentic curriculum (Sobel 1994). One avenue of access into authentic curriculum is to use children's interest in special places to fuel well-designed mapping projects. The following portrait is a beautiful example of wedding heart and mind through careful curriculum design.

The description and the following student worksheets are based on work done in Terry Monette's science classes at the Harvard Elementary School in Harvard, Massachusetts. Maggie Stier, then an intern and currently the director of Friends of the John Hay Estate in Newbury, New Hampshire, helped to develop many parts of the original unit.

MAPPING SPECIAL PLACES

In preparation for an Earth Day celebration, Maggie decided to initiate a Special Places mapping unit with the ninety fifth-grade students in four classes at the Harvard Elementary School. With her supervising teacher, Terry Monette, she decided on two preparatory activities that would lead into the major assignment. First, since the teachers were going to have the children work at home, where they would need to measure large areas, they decided to teach the children to calculate the length of their own pace. This way, each child would have a handy, low-tech measuring device. Second, they wanted to have the children do a school-based map-reading activity to make sure that everyone had basic competence with map-reading skills.

The "Determining Your Pace for Mapmaking" activity was conducted on the soccer field at the school. First, the teachers placed sets of cones one hundred feet apart and had the children count the number of paces it took them to walk one hundred feet. A pace is considered to be a complete left and right cycle, so you count a pace by counting each time your right foot hits the ground. The children paced the one-hundred-foot distance three times and then figured out the length of their pace by using calculators to do the mathematics. (Please see Worksheet #1 for specific directions.)

Then, to test the children's ability to use their new measuring device, the teachers laid out four markers and asked the children to calculate the distance between #1 and #2, between #2 and #3, and so on. The children then matched their calculations against the correct measured distances, which were kept secret until the end of the activity. (Please see Worksheet #2 for specific directions.)

In order to provide the children with opportunities to practice map reading and to engage the children in school-based nature observations, the teachers conducted a two-day activity focused on observing erosion. On the first day, Maggie brought a stream table into the classroom and filled it at different times with sand, gravel, loam, clumps of grass, and sticks so that the children could play around with how the forces of

---

Worksheet #1
Determining Your Pace for Mapmaking

Name: _____   **Date:** _____

1. Count the number of paces it takes you to walk 100 feet. Begin with your left foot and count every time your right foot strikes the ground. Pace the distance three times and record your results. (2 steps = 1 pace)

1st trial _____
2nd trial _____
3rd trial _____

2. Find the average number of paces that you take to walk 100 feet. To do this, add the three trials together and divide by three. Show your work in the space above and then record your answer below.

The average number of paces it takes me to walk 100 feet is _____.

3. To determine the length of your pace, divide 100 by the average number of paces it takes you to walk 100 feet. Do your work here.

The length of my pace is _____.

4. Round off the average length of your pace to the nearest foot: _____.

5. You can now use this length to get an approximate measurement of distance by "pacing the distance." Do the problems on the back of this page. Now do the problems on the next page.

Determine the approximate distance between the markers by using your pace. (Multiply the number of paces you took times the average length of your pace.)

Worksheet #1:
Determining Your Pace
for Mapmaking

running water affect the earth's surface and see how erosion might be controlled through man-made structures. On the second day, Maggie drew a map of the school and school yard and created descriptions of four erosion areas that she wanted the children to find on the school grounds. Maggie described the results of this activity:

> When we went outside to observe erosion, it was raining, which was good for our observations but tough on the worksheets-and-clipboard method of recording data. We dried the sheets on the radiator. The next day the kids all told me they had had a great

Worksheet #2: Pacing the Basketball Court

Worksheet #2
Pacing the Basketball Court

The markers on the outside of the basketball court identify the boundaries of this "Special Place." Draw a sketch of the boundaries. Then determine the approximate distance between the markers by using your pace, and mark this on your sketch. (Multiply the number of paces you took times the average length of your pace.)

#1 to #2
Number of paces _____ X _____(the average length of your pace) = _____ feet (approximate distance between #1 and #2)

Approximate distance between #1 and #2 is _____ feet.

#2 to #3
Number of paces _____ X_____(the average length of your pace) = _____ feet (approximate distance between #2 and #3)

#3 to #4
Number of paces _____ X _____(the average length of your pace) = _____ feet (approximate distance between #3 and #4)

#4 to #1
Number of paces _____ X _____(the average length of your pace) = _____ feet (approximate distance between #4 and #1)

Add one prominent feature to your sketch and determine the distance from one of the boundaries.

time! All this, while not directly related to the process of making maps, served a valuable introductory function. It got kids to read maps, and it got them focused on looking closely at things, asking questions, and trying to figure out answers.

With these preparatory activities successfully completed, the teachers then introduced the major assignment. Each child was given two weeks to complete a map of a special place somewhere in the vicinity of home. To open up children's notions of graphic possibilities, the teachers showed the children many different map examples—the map of the school grounds from the worksheet, the sketch map Maggie used in order to gather information before she made the map, the map of the school's nature trail from the guide to local conservation lands, a USGS topographical map, and a historic watercolor map of the local Shaker village. Finally, Terry displayed a map of her own special place made as part of a mapmaking course she had taken. Showing the children her own map was a major contributing factor to this project's success.

The teachers explained that the reason for making this assignment was to encourage children to look closely at one small piece of the natural environment, preferably a piece of the world that they felt strongly about. If the children didn't have a special place, the teachers required them to find one and to become more closely connected to it. Having a special place, they suggested, was valuable for knowing one place really well, having a place to be by yourself, and having a place to retreat to when you are upset.

Maggie and Terry designed packets for each child to complete as a way of scaffolding the activity. Since the work had to be done mostly at home, the packet was essential in keeping the children focused on the task. (See Worksheets #3 and #4.) The packet asked the students to describe the location of the place and its prominent natural features, tell why the place was special to them, make at least two visits and indicate when these occurred, make observations of natural phenomena and sounds, draw something they saw at the site, and make an initial sketch map of the area. These packets were checked in class to make sure each child was making progress, and then the final project guidelines were distributed. (See Worksheet #6: Special Places Map Project.) The guidelines specified the features that needed to be included for the final map:

- Size: 22-by-28 inches on poster board
- Title
- A statement about why this place is special to you
- Map key and/or labels
- Distance scale
- Color
- Your name
- Two or more of the following:
  - A detailed wildlife or plant sketch
  - A drawing of the view from your special place

- A poem based on your observations of nature or sounds
- Dated observations
- Specimens collected at the site and used as a border or to mark some important features on your map
- Photographs
- An investigation of a question about the place

This checklist provided the framework for the final evaluation. (See Special Places Map Assessment.) The students were given two weeks to complete the map poster, including April vacation, predictably a nice time

---

Worksheet #3
Mapping Your Special Place

Name: _____   **Date:** _____

Use the method of measurement you practiced on the basketball court at school to help you make a map of your Special Place.

Make a working sketch of your Special Place in the space below, showing the boundaries of your Special Place, its approximate size, and at least three prominent features in their proper location. Use your pace to help you measure distances. Mark these distances on your sketch.

My pace is _____ feet.

for children to be out exploring the woods. The due date was a few days before the school's Earth Day celebration so the posters could be displayed at the fair.

For the children, the hardest part of creating the final map was measuring and drawing to scale. When the teachers realized that many of the children were struggling with this part of the project, they drew up supplemental directions and reviewed them in science class (see Worksheet #5, "Drawing a Map to Scale"). Even so, they agreed that more guided experiential activities like the "Measuring Your Pace" activity would have helped many of the students to solidify this skill.

When the due date arrived, Maggie and Terry were astounded that all ninety students had their projects completed on time. The Earth Day fair turned out to be a powerful motivator for getting the assignments done. Kids who were sick even asked their parents to make a special trip to drop off the maps.

On the day the maps came in, Maggie conducted a sharing time with each class so that the children could describe their special places to their classmates. She noted, "Though some were shy about reading personal statements or poems, all were encouraged to do so, and it was a wonderfully engaged and excited group of kids."

The poems and commentary on the posters are testimony to the effectiveness of this mapping project. The assignment helped the children to articulate both the sense of immersion that they felt in the natural world and the emotional solace that these places can provide. One child's poem portrays a Zen-like attentiveness:

My Place The Peaceful Province

Silence still
Then suddenly
Alive!
You can hear the birds
Chirping to welcome spring.
You can hear the wind
Rustling through the trees
To make its destination
Before Mother Earth
Calms it.

You can see the salamander
Perched on a rock
Listening intently
Silence still.

Many children appreciate the opportunity to escape the confusion and turmoil of everyday life and be soothed by nature in their special places. One boy wrote,

When I am mad, I come here to be alone.
Sometimes, when I am sad, I watch the trees
Swaying in the wind.
When I am lonely the bird calls and trees
Make me feel at home.
When I am mad, I come here to be with them.

Worksheet #4:
Special Places Project

Worksheet #4
Special Places Project

Name: _____   **Date:** _____

This Special Places project is an opportunity to appreciate and learn more about an outdoor place that is special to you. Whether it is threatened by overuse or pollution, preserved as conservation land, or just part of your backyard, you will be asked to look closely at nature there.

You will record your observations at various times during the year. Your careful observations will help you decide on something specific you want to investigate about your special place. Later in the year you will learn some skills that will help you create a map of your Special Place. (Separate directions for the map will be handed out then.)

A. Where is your special place located? (Give town, street, name of property owner or conservation land)

_____

_____

_____

_____

B. Describe the prominent natural features of this landscape: (Woods, fields, rocks, stream, pond or river, pathways, hills, etc.)

_____

_____

_____

_____

C. Why is this place special to you?

_____

_____

_____

_____

D. Spend at least fifteen minutes in your special place on two different days:

Visit #1 _____
       (date)   (time of day)   (weather)

Visit #2 _____
       (date)   (time of day)   (weather)

This assignment struck the perfect balance between science and art, between thinking and feeling. The clear mathematical and scientific directions implied that this was an important and serious task. The emphasis on personal expression, attentiveness, and artistry invited children to express the heartfelt connections they feel with the natural world. The result was a true expression of a child's sense of place, geography claimed by feelings.

E. During your visits, record your observations of the following:
Birds: _____
Mammals: _____
Insects: _____
Aquatic (Water) Life: _____
Trees: _____
Other Plants: _____
Evidence of Animals: (homes, droppings, exoskeletons, etc.) ____
_____
_____

F. Close your eyes and listen to the sounds of your special place for several minutes. Write a description of those sounds and try to figure out what you were hearing.

_____
_____
_____

G. On the back of this page, make a list of things that interest you about your special place that you would like to know more about. Use your observations to help you.

When the children were asked what they liked about the project, some comments were:

I liked being outdoors to do homework.
I liked having to visit my special place.
It felt good to find a special place.
I liked seeing the good finished projects.
I appreciated the freedom to do what we wanted for the project.

Worksheet #5:
Drawing a Map to Scale

Worksheet #5
Drawing a Map to Scale

Drawing a map to scale means that distances on the map are shown in the same proportions as they exist in reality. That is, if two trees are ten feet apart in your yard, you need to show that they are ten feet apart on your paper. Since you probably don't have a piece of paper that is ten feet long, you have to make distances on your paper stand for, or *represent*, the real distances in the woods. It's like having a little black square on a map to *represent* a house.

So let's say you choose your scale to be 1 inch = 1 foot. If your trees are ten feet apart in your special place, then on your map you would show them as ten inches apart. On your map you would say:

scale: 1 inch = 1 foot

The tricky part about scale is that it will depend on two things—the size of the piece of paper for your map AND the size of the area you want to map. If you're mapping your town, you will probably use a different scale than if you're mapping your backyard. Here's how it works.

1. Measure your poster board. All of your map must fit within this size. (Plan out where you will put the title and other required items.)

2. Find your longest boundary. Use the working sketch from Worksheet #3: Mapping Your Special Place to help you determine the scale of your map. You should have each boundary labeled with its approximate distance. Find the boundary with the longest distance. Draw this boundary line on the poster board as long as you can make it without going off the paper. Measure this distance with a ruler. If the longest boundary line you can draw on your map is twenty-five inches and the longest distance in real space is fifteen feet, then your ratio is:

25 inches = 15 feet

3. Calculate your scale. Reduce the fraction by dividing each side by five to get:

scale: 5 inches = 3 feet.

© 1998 by David Sobel from *Mapmaking with Children*. Portsmouth, NH: Heinemann

A horde of volunteers set up the display in the gym overnight. The fifth-grade maps, displayed on one wall, were the focal point of the school-wide celebration, which was titled "Here's Looking at You, Harvard." Maggie and Terry displayed an enlarged town map on which the children placed numbered stickers that corresponded to the numbered stickers on their maps. Since there were ninety special places, the stickers speckled the enlarged map, and the children's maps filled up a substantial amount

4. Make the scale even. If you come up with odd numbers (like 23 inches = 13 feet), then you should probably do a little adjusting so your scale is more even. This makes the math you have to do much easier. In this case, you should probably make your map a little smaller so it's twenty inches, and then add a foot or two to your area to make it fourteen or fifteen feet. Then your ratio could be:

20 inches = 14 feet for a scale of: 10 inches = 7 feet

or

20 inches = 15 feet for a scale of: 4 inches = 3 feet

Another choice would be to really increase or decrease the size of your map to get an easier scale. You could decrease it so:

15 inches = 15 feet for a scale of 1 inch = 1 foot

Or you could increase it so that:

30 inches = 15 feet for a scale of 2 inches = 1 foot

5. Use your scale to place other features on your map. Let's say your scale is 5 feet = 3 inches

a. If the distance between the stone wall and the berry bush is ten feet, then you calculate the distance on the map:

2 X 5 feet = 10 feet so 2 X 3 inches = 6 inches

b. If the distance between the stone wall and the berry bush is eight feet, then you calculate the distance on the map:

8/5 X 5 feet = 8 feet so 8/5 X 3 inches = 24/5 or 4 4/5 inches

Since rulers don't usually measure in fifths, the easy thing is to round off four-fifths of an inch to three-fourths of an inch.

6. A word to the wise. Drawing your map to scale will be easiest if you figure out an easy scale in the beginning. The challenge is coming up with an easy scale but not making your map too big or too small. If your map is too big, it's harder to store. If your map is too small, it's harder to see things on it. Figuring out the scale for your map is a good lesson in learning how to compromise.

of wall space. It was dramatically apparent that the children had a deep affinity for their local natural places. What a wonderful way to convey to parents, town officials, developers, and local businesspeople that the preservation of local open space needs to be a community priority. (See Figures 8–1 and 8–2.)

Everyone who looked at the maps was impressed with how uniformly beautiful and well designed they were. Maggie attributed this to a variety

---

Worksheet #6
Special Places Map Project

Name: _____   **Date:** _____

1. Use the work you have done during the year to make the final map of your special place. This should be done on poster board or similar-sized paper (22 by 28 inches). Remember that your poster will include your map, a title, and a sketch or poem or whatever extra things you decide to include.

2. Use the method you have learned in school for "pacing" to map your Special Place. This will give you the approximate distances of the boundary lines of your Special Place. Then calculate the ratio and proportion of your Special Place as we did in the "Drawing a Map to Scale" activity. Use this information to help you draw your final map to scale on your poster board.

3. The map must include the following:

_____ a. Title
_____ b. Important landscape features
_____ c. A statement about why this place is special to you
_____ d. Map key or labeled features
_____ e. Distance scale. The scale will depend on the size of the area you are mapping. (For example: 1 inch = 3 feet)
_____ f. Colored or shaded feature
_____ g. Name (clearly visible on front)
_____ h. Include two or more of the following:
   * A detailed wildlife or plant sketch
   * A drawing of the view from your place
   * A poem based on your observations of nature or sounds
   * Specimens collected from your Special Place and used as a border or to mark some important features on your map

Remember to be creative and have fun! If you use watercolors to paint your map, you might want to use water collected from your Special Place to mix the paint.

Make sure your map is neat and carefully done. Be creative and have fun!

Special Place map due: _____
Special Place maps will be on display at the Earth Day Fair on _____

of factors. One factor was Terry's sample map—an elegantly simple model. From her house, a path led across the railroad tracks, through pine woods, and down to a river. The railroad tracks were made with toothpicks, and the pine trees were made out of cut paper. The border around her map was made from leaves, pressed flowers, feathers, and pieces of mica found in her woods.

Special Places Map Assessment

Name: _____   **Date:** _____

Communication
4. Map is visually attractive and organized, and educates and informs the viewer about this Special Place in an impressive way.
3. Map is visually attractive and organized, and educates and informs the viewer about this Special Place.
2. Lack of neatness, organization, or information detracts from educating and informing the viewer about this Special Place.
1. Lack of neatness, organization, or information interferes with the viewer's understanding of this Special Place.
0. The assignment was late or not turned in.

Following Directions
4. All requirements have been met. (Use checklist from direction sheet.)
3. Most requirements have been met.
2. Some requirements have been met.
1. Few requirements have been met.
0. The assignment was late or not turned in.

Proofreading
4. There are no mechanical errors.
3. There are few mechanical errors.
2. There are one or two significant mechanical errors (e.g., word in title misspelled).
1. There are many mechanical errors.
0. The assignment was late or not turned in.

Teacher Comments:

Maggie had been hesitant to show the children Terry's refined map because she didn't want them to copy it. She also didn't want to intimidate them with Terry's sophisticated work. Instead, it appeared that just the opposite had happened. One or two of the children made uninspired facsimiles of the teacher's map, but the rest of the children followed their own muse, sure of the standard that had been set. Terry's attention to detail, clean design, and collage elements, and her sincere expression of love for place, encouraged the children to really invest themselves in their maps. Rather than copying the teacher's work, the children picked up on some of her techniques and developed them in ingenious ways.

Too often, I think, progressive educators avoid having the teachers demonstrate their own artistry because they fear it will steal away the children's creativity. I believe the opposite is true. My conviction is that the teacher's artistry inspires or compels the children to do their best work. If teachers make beautiful maps, children will too.

A second factor contributing to the project's success was its developmental appropriateness. Reflecting on the project, Maggie said,

> I was astounded at how perfect it was for fifth graders. They were able to pull together so many different strands of knowledge and experience into real works of art. Their conceptual development spans a range, but their variety of abilities to depict a place in two dimensions was part of what made it so interesting a challenge. Many did bird's eye views that covered a wide area. Others did relatively small landscapes in a perspective view. A few still strug-

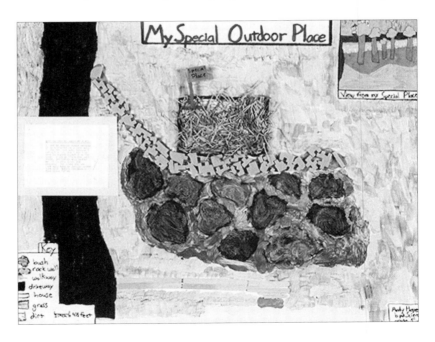

FIGURE 8–1
Special place map drawn by a fifth-grade student in Harvard, Massachusetts.

gled with a single-point perspective and had elements folded down flat around a central imaginary viewpoint. Everyone was proud of their finished product, though. This assignment gave them ownership of their learning and allowed them to choose a mode of expression that suited them. School should always be so much fun and so interesting.

Despite the fact that they were all fine works of art, the teachers did assess children on their work. The evaluation rubric was based on the Special Places map directions that had been distributed to the children (see Special Places Map Assessment). If children followed the directions, spelled correctly, and worked with neatness and care, they did well. Many children did well, which pleased the teachers. The hard part was giving children lower grades because they had forgotten essentials such as name, title, scale, or optional element—especially when they had created a decent map or landscape view. But, they had neglected to follow the directions.

Terry has continued to use this project for the past two years, and it's still a hit with the fifth graders. It's been stretched out over almost half of the school year so that children really have the opportunity to become immersed in their places and the mapping process. (This is why some of the worksheets refer to a yearlong process.) The assessment includes a quantitative component that measures the technical aspects, and Terry includes a narrative assessment as well. For assignments like this, with a

FIGURE 8–2
Special place map drawn by a fifth-grade student in Harvard, Massachusetts. This student requested permission to map a favorite family vacation spot in Chester, Maryland.

strong personal component, a narrative assessment allows the teacher to praise a child for good work even when some of the mechanics are missing.

What happens to these special-places maps? After being displayed in the school or at the public library in the summer, they are returned to the children. Some live on bedroom walls or on the front of the refrigerator for a few weeks; others get lost in the clutter under the bed. A few may be framed by conscientious parents to be hung in the den; others will get thrown away. These kinds of big projects are hard to store for posterity. I suspect, however, that most of them will live on in the children's minds.

One of my strongest memories from elementary school is of the book report I did on *Born Free*, the story of Elsa the lion. Instead of demanding the run-of-the-mill written book report, the teacher provided the option of creating a diorama of a scene from the book. I tilted a PF Fliers shoe box on its side and created a swath of savanna inside. I remember the cottony clouds, the bushes secured in modeling clay, the waterhole, Elsa on a rock. It was one of the few opportunities I had to create an artistic representation of something that was important to me. For the children in Terry's class, their special places have been elevated to significant memories by sensitive, authentic curriculum.

BIBLIOGRAPHIES

WORKS CITED

ADAMSON, JOY. 1987. *Born Free: A Lioness of Two Worlds.* New York: Pantheon.

ALEXANDER, LLOYD. 1969. *The Chronicles of Prydain.* New York: Yearling Books.

ARBIB, ROBERT. 1971. *The Lord's Woods.* New York: W. W. Norton & Co.

BARTH, BETH. 1985. *Mapping: ESS Teacher's Guide.* Nashua, NH: Delta Education.

CARLSON, NATALIE SAVAGE. 1958. *Family Under the Bridge.* New York: HarperCrest.

CATLING, SIMON. 1985. *Mapstart 1, 2, & 3.* Essex, England: Collins-Longman Group Ltd.

CLINTON, HILLARY. 1996. *It Takes a Village to Raise a Child.* New York: Simon & Schuster.

COBB, EDITH. 1959. "The Ecology of Imagination in Childhood." In *Daedalus.* Boston: American Academy of Arts and Sciences.

DILLARD, ANNIE. 1987. *An American Childhood.* New York: Harper & Row.

DRISCOLL, ROSLYNN. 1992. "The Place of Imagination." *Places* (winter).

GANNETT, RUTH STILES. 1987a. *The Dragons of Blueland.* New York: Knopf.

——. 1987b. *Elmer and the Dragon.* New York: Knopf.

——. 1987c. *My Father's Dragon.* New York: Knopf.

GARDNER, HOWARD. 1993a. *Multiple Intelligences: The Theory in Practice.* New York: Basic Books.

——. 1993b. *The Unschooled Mind.* New York: Basic Books.

GEORGE, JEAN CRAIGHEAD. 1959. *My Side of the Mountain.* New York: E. P. Dutton.

HART, ROGER A. 1979. *Children's Experience of Place.* New York: Irvington Publishers, Inc.

Hillary, Sir Edmund. 1962. *High in the Thin, Cold Air.* New York: Doubleday.

Kallett, Tony. 1995. "Homo Cartographicus." In *Few Adults Crawl: Thoughts on Young Children Learning.* North Dakota Study Group on Evaluation. Grand Forks, ND: University of North Dakota.

Lindgren, Astrid. 1985. *Ronia, the Robber's Daughter.* New York: Viking.

Millstone, David. 1989. "Adventures in Our Backyard." In *Great Hike Downstream Newsletter.* Norwich, VT: Marion Cross School.

Milne, A. A. 1926. *Winnie-the-Pooh.* New York: Dutton.

Mitchell, Lucy Sprague. 1991. *Young Geographers.* New York: Bank Street College of Education.

Murrow, Liza Ketchum. 1991. *The Ghost of Lost Island.* New York: Holiday House.

Norman, Howard. 1988. *Northern Lights.* New York: Pocket Books.

O'Dell, Scott. 1960. *Island of the Blue Dolphins.* Boston: Houghton Mifflin.

Orr, David. 1994. *Earth in Mind: On Education, Environment and the Human Prospect.* Washington, DC: Island Press.

Parish Maps: Celebrating and Looking After Your Place. 1997. London: Common Ground.

Pearce, Joseph Chilton. 1992. *Magical Child.* New York: E. P. Dutton.

Peet, Bill. 1985. *The Kweeks of Kookatumdee.* Boston: Houghton Mifflin.

Ransome, Arthur. 1984. *Secret Water.* New York: Puffin Books.

Sobel, David. 1993. *Children's Special Places.* Tucson, AZ: Zephyr Press.

———. 1994. "Authentic Curriculum." *Holistic Education Review* 7 (2).

Spier, Peter. 1961. *The Fox Went Out on a Chilly Night.* New York: Doubleday.

Stafford, Kim. 1986. "The Separate Hearth." In *Having Everything Right: Essays of Place.* Lewiston, ID: Confluence Press.

Steiner, Jorg. 1977. *Rabbit Island.* New York: Bergh.

Stevenson, Robert Louis. [1883], 1997. *Treasure Island.* Austin, TX: Holt, Rinehart & Winston, Inc.

Tolkien, J. R. R. 1966. *The Hobbit.* Boston: Houghton Mifflin.

Vygotsky, Lev. 1980. *Mind and Society.* Cambridge, MA: Harvard University Press.

Wood, Dennis. 1992. *The Power of Maps.* New York: Guildford Press.

Yolen, Jane. 1987. *Owl Moon.* New York: Philomel.

RESOURCES FOR TEACHERS

The following list of books contains some of my favorite resource books. I have included books that provide expert guidance on mapmaking with children, books that articulate the connection between mapmaking and developing a sense of place, and a few children's books in which mapmaking is the core topic of the story. The bibliography is selective rather than exhaustive. Some of my personal copies of these books are dog-eared and spineless. I have used them extensively in my teaching, and you will find the imprint of their ideas in the pages of this book. My thanks to these authors who documented their fine work with children.

ABERLY, DOUG, ed. 1993. *Boundaries of Home: Mapping for Local Empowerment.* Philadelphia, PA: New Society Publishers. The philosophical framework for bioregional mapping. Inspiring examples of community mapping projects that create a commitment to place.

BARTH, BETH. 1985. *Mapping: ESS Teacher's Guide.* Nashua, NH: Delta Education. During the science curriculum revolution of the 1960s, the Elementary Science Study (ESS) curriculum kits and guides were the hit of the decade. As with lots of kit programs, many of these have fallen into disuse, but the Teacher's Guides are still one of the best elementary science resources around. Many teachers revere the Mapping Teacher's Guide. The guide illustrates diverse ways to get children involved in making classroom, neighborhood, and topographical maps. The photographs of children at work and the pictures of children's maps make you think, My students could do that. Specific techniques for doing map and compass work, creating contour maps, and surveying the playground are all clearly explained. I am indebted to this author for some of the ideas in this book.

BELL, NEILL. 1982. *The Book of Where.* Boston: Little, Brown. A child-friendly approach to geography education from the same people who brought you *The Backyard History Book* and the *I Hate Math! Book.* Though the focus is more on geography activities than on mapmaking, the funny drawings, jokes, and hands-on activities make this an appealing and useful book.

BESANT, PAM, AND ALAISTAIR SMITH. 1993. *How to Draw Maps and Charts.* Tulsa, OK: Usborne/EDC Publishing. One of the How to Draw Series with very useful, well-illustrated guidelines for mapping projects.

CATLING, SIMON. 1985. *Mapstart 1, 2 & 3.* Essex, England: Collins-Longman Group Ltd. While geography curricula have been in hibernation in the United States, they have been running rampant in the United Kingdom. One of the best is Mapstart, a "series of graded map skill and atlas books and related materials [that] gradually develop the basic skills children need to be able to understand and actively use plans and maps of all kinds." There are three books in the series and each is accompanied by an activity guide. Though the focus is on map reading rather than on mapmaking, this is really an elegantly designed set of learning materials. One of the ingenious features of the first book is a step-by-step progression from a side view of a three-dimensional Lego village to top-down photos of the village to a symbolic map of the village. The illustrated transition from concrete to abstract is elegant. Though all the content is British, these could be exceedingly valuable in elementary curricula in the United States.

CHAPMAN, GILLIAN, and PAM ROBSON. 1993. *Maps and Mazes.* Brookfield, CT: The Millbrook Press. Very attractive illustrations showing some ingenious elementary classroom projects. Demonstrates mapmaking with a variety of artistic materials that result in simple but beautiful maps.

Discovering Your Life-Place, A First Bioregional Workbook. 1995. San Francisco: Planet Drum Publications. A handy workbook for helping students create first maps of their communities and bioregions.

DISLEY, JOHN. 1971. *Map and Compass—Orienteering.* Burlington, Ontario: Orienteering Services of Canada. This guide is a bit antiquated, but it provides a great introduction to map and compass use with students. Step-by-step activities introduce map reading and compass use until all the pieces are put together for doing simple orienteering with students.

HARRIS, MELVILLE. 1972. *Starting from Maps.* School Council Environmental Studies Project. London: Rupert Hart-Davis Educational Publications. One of the best guides to community studies through mapmaking from teachers in the United Kingdom. This book is old and out of print, but it's worth searching for.

HARTMAN, GAIL. 1991. *As the Crow Flies: A First Book of Maps.* New York: Simon & Schuster. A developmentally appropriate primer on understanding maps for very young children.

INNER LONDON EDUCATION AUTHORITY. 1981. *The Study of Places in the Primary School.* London: Inner London Education Authority. The most sophisticated portrait of place-based education that I have ever found. Complex charts articulate mapmaking and map reading skills at different developmental levels. Lavish illustrations of children's geography work throughout the elementary grades. Out of print and hard to find, but worth the search.

INSTITUTE FOR COMMUNITY ENVIRONMENTAL MANAGEMENT. 1996. *Valley Quest: Quest Maps and Teacher's Guide.* Keene, NH: Antioch New England Graduate School. Two different publications describing an integrated curriculum approach to creating educational treasure hunts. Students study local history and natural science and then create treasure hunts to lead visitors to special community spots. A unique place-based education mapmaking project. The collection of classroom maps and the teacher's guide are available separately.

MITCHELL, LUCY SPRAGUE. 1991. *Young Geographers.* New York: Bank Street College of Education. The first and still one of the best books on geography education in the elementary grades. The author articulates a clear developmental model for engaging students in geography and clearly explains a variety of different mapmaking techniques. One of the classics in the field.

Parish Maps: Celebrating and Looking After Your Place. 1997. London: Common Ground. A beautiful little how-to guide on one of the best sense-of-place mapping projects in the past two decades. Lots of maps created by community artists and elementary classrooms that demonstrate both geographical accuracy and an appreciation of local places.

RANSOME, ARTHUR. 1984. *Secret Water.* New York: Puffin Books. Ransome is famous for his Swallows and Amazons series of books for British children. In this one, five brothers and sisters are marooned by their father on a marshy island with "a blank map [of the surrounding archipelago] that doesn't do more than roughly show what's water and what isn't." Their challenge is to fill in all of the details of the map over the following week. The children's explorations are described in the text and shown in a series of progressively more detailed maps. This book could provide a great stimulus and framework for local mapping projects around the school.

ROGER TORY PETERSON INSTITUTE OF NATURAL HISTORY. 1996. *The Selborne Celebration: Annual Journal of the Selborne Project.* Jamestown, NY: Roger Tory Peterson Institute of Natural History. The annual publication of this unusual project in western New York and northwestern Pennsylvania. Neighborhood maps, architecture, and natural history studies by middle school students focusing on one square kilometer around the school.

SALT SPRING ISLAND COMMUNITY SERVICES. 1996. *Giving the Land a Voice.* Salt Spring Island, British Columbia: Salt Spring Island Community Services. A beautiful and thought-provoking book about developing a sense of place through mapmaking. Numerous Canadian and Native American examples.

SOBEL, DAVID, ed. 1989. "Maps and Mapmaking." *CONNECT: The Newsletter of Practical Science and Math for K–8 Teachers* (May). Brattleboro, VT: The Teacher's Laboratory. Descriptions of classroom-based projects on mapmaking with children.

TAYLOR, BARBARA. 1993. *Maps and Mapping.* New York: Kingfisher. Highly graphic and beautifully illustrated. Good activities on mapmaking and map reading.

U.S. GEOLOGICAL SURVEY. 1996. *Map Adventures; What Do Maps Show?; Exploring Maps.* Reston, VA: U.S. Geological Survey. Developmentally sequenced learning packets for teachers. The focus is on map reading rather than on mapmaking, but the maps are beautiful. These packets are a fine complement to a community-based geography program.

WATTS, STEVE. 1996. *Make It Work: Maps—The Hands on Approach to Geography.* Chicago: World Book/Two Can Publishing. One of the best new books on hands-on geography techniques with children. Elegantly illustrated with developmentally appropriate graphics explaining mapmaking techniques.

WENTWORTH, D. F., et al. 1972. *Mapping Small Places: Examining Your Environment.* Toronto: Holt, Rinehart & Winston of Canada. Examining Your Environment was one of the best environmental education curriculum series available. Regrettably, most of the titles are now unavailable. Other titles in the series that provide good material for mapping are *Mini-Climates, Ecology in Your Community,* and *The Dandelion. Mapping Small Places* is undoubtedly one of the best curriculum guides to mapmaking with children ever printed. The guide is lavishly illustrated with four-color photographs. The pictures of children at different stages in the mapmaking process make the translation to your classroom highly doable. Above and beyond this technical clarity, the guide suggests some wonderful projects. Children pursue questions such as, How do the size and shape of an anthill change from day to day? or How can the cross-section of a snowdrift be mapped? If you find this book, scoop it up. Nothing of comparable usefulness has been published in the last twenty years. I am indebted to these authors for some of the ideas in this book.

WHEATLEY, NADIA, and DONNA RAWLINS. 1987. *My Place.* Long Beach, CA: Australia in Print Publishers. An unusual children's book that traces the history of Australia through maps of the same neighborhood over two hundred years. The maps are detailed and compelling and hold many fascinating stories. A great model for a sophisticated community studies project in the upper elementary grades.

CHILDREN'S BOOKS WITH MAPS

All of these books contain classic, child-appropriate maps on the endpaper or in the text.

GAMMELL, STEPHEN. 1983. *Git Along, Old Scudder.* New York: Lothrop, Lee & Shepard.

GANNETT, RUTH STILES. 1948. *My Father's Dragon.* New York: Random House.

———. [1951], 1979. *The Dragons of Blueland.* New York: Knopf.

HOLLING, HOLLING CLANCY. 1941. *Paddle-to-the-Sea.* Boston: Houghton Mifflin.

LEIGH, SUSANNAH. 1990. *Puzzle Island.* Newton, MA: Educational Development Center, Inc.

MICHL, REINHARD. 1985. *A Day on the River.* Hauppauge, NY: Barron's Educational Series, Inc.

MILNE, A. A. 1926. *Winnie-the-Pooh.* New York: Dutton.

MURROW, LIZA KETCHUM. 1991. *The Ghost of Lost Island.* New York: Holiday House.

ROCKWELL, ANNE. 1994. *The Way to Captain Yankee's.* New York: Macmillan.

STEVENSON, ROBERT LOUIS. [1883], 1997. *Treasure Island.* Austin, TX: Holt, Rinehart & Winston, Inc.

TOLKIEN, J. R. R. 1966. *The Hobbit.* Boston: Houghton Mifflin.

WHEATLEY, NADIA, and DONNA RAWLINS. 1987. *My Place.* Long Beach, CA: Australia in Print Publishers.

WILLIAMS, VERA B. 1981. *Three Days on a River in a Red Canoe.* New York: Greenwillow.

BOOKS WITH PANORAMIC VIEWS

The illustrations in these books contain panoramic views, and are easily accessible to children. These views serve as good models for maps, and they're good starting points for making more aerial view maps.

BROWN, RUTH. 1987. *Our Puppy's Vacation.* New York: Dutton.

BURNINGHAM, JOHN. 1967. *Harquin, The Fox who went down to the Valley.* Old Tappan, NJ: Bobbs-Merrill.

COONEY, BARBARA. 1982. *Miss Rumphius.* New York: Viking.

HELLDORFER, MARY CLAIRE. 1991a. *The Mapmaker's Daughter.* New York: Bradbury Press.

———. 1991b. *Sailing to the Sea.* New York: Viking.

LOCKER, THOMAS. 1984. *Where the River Begins.* New York: Dial Books.

MACLACHLAN, PATRICIA. 1994. *All the Places to Love.* New York: Harper-Collins.

MUNRO, ROXIE. 1985. *The Inside-Outside Book of New York City.* New York: Dodd, Mead.

SPIER, PETER. 1961. *The Fox Went Out on a Chilly Night.* New York: Doubleday.

STEINER, JORG. 1977. *Rabbit Island.* New York: Bergh.

WINTER, JEANETTE. 1988. *Follow the Drinking Gourd.* New York: Knopf.

YOUNG, RUTH. 1991. *Daisy's Taxi.* New York: Orchard.

BOOKS FOR MAPPING

These stories all move through a landscape that is illustrated sequentially. These books provide good challenges for creating a comprehensive map that does not exist in the text.

ALEXANDER, LLOYD. 1969. *The Chronicles of Prydain.* New York: Yearling Books.

ARNOLD, TEDD. 1993. *Green Wilma.* New York: Dial.

BRETT, JAN. 1992. *The Trouble with Trolls.* New York: Putnam.

BROWN, RUTH. 1987. *Our Puppy's Vacation.* New York: Dutton.

BURNINGHAM, JOHN. 1967. *Harquin, The Fox who went down to the Valley.* New York: Bobbs-Merrill.

BURSTON, PATRICK. 1985. *The Jungle of Peril.* New York: Simon & Schuster.

CARLSON, NATALIE SAVAGE. 1958. *Family Under the Bridge.* New York: HarperCrest.

CARRICK, DONALD. 1988. *Harold and the Great Stag.* New York: Clarion Books.

COONEY, BARBARA. 1982. *Miss Rumphius.* New York: Viking.

HELLDORFER, MARY CLAIRE. 1991. *Sailing to the Sea.* New York: Viking.

LINDBERGH, REEVE. 1987. *The Midnight Farm.* New York: Dial.

LINDGREN, ASTRID. 1985. *Ronia, the Robber's Daughter.* New York: Viking.

LOCKER, THOMAS. 1984. *Where the River Begins.* New York: Dial Books.

MCCLOSKEY, ROBERT. [1941], 1976. *Make Way for Ducklings.* New York: Puffin.

———. [1948], 1976. *Blueberries for Sal.* New York: Puffin.

MCLEOD, EMILIE WARREN. 1975. *The Bear's Bicycle.* Boston: Little, Brown.

MARTIN, CHARLES E. 1983. *Dunkel Takes a Walk.* New York: Greenwillow.

O'DELL, SCOTT. 1960. *Island of the Blue Dolphins.* Boston: Houghton Mifflin.

OAKLEY, GRAHAM. 1981. *Hetty and Harriet.* New York: Atheneum.

———. 1982. *The Church Mice in Action.* New York: Atheneum.

———. 1984. *The Church Mice Spread Their Wings.* New York: Atheneum.

PEET, BILL. 1985. *The Kweeks of Kookatumdee.* Boston: Houghton Mifflin.

ROBERTS, TOM, adap. 1995. *Red Riding Hood.* New York: Simon & Schuster.

SELDEN, GEORGE. 1983. *Chester Cricket's Pigeon Ride.* New York: Bantam, Doubleday, Dell.

SPIER, PETER. 1961. *The Fox Went Out on a Chilly Night.* New York: Doubleday.

VAN ALLSBURG, CHRIS. 1979. *The Garden of Abdul Gasazi.* Boston: Houghton Mifflin.

———. 1985. *The Polar Express.* Boston: Houghton Mifflin.

YOLEN, JANE. 1987. *Owl Moon.* New York: Philomel.